Artificial Intelligence and Cybersecurity

Green Engineering and Technology: Concepts and Applications

Series Editors: *Brujo Kishore Mishra, GIET University, India and Raghvendra Kumar, LNCT College, India*

Environment is an important issue these days for the whole world. Different strategies and technologies are used to save the environment. Technology is the application of knowledge to practical requirements. Green technologies encompass various aspects of technology which help us reduce the human impact on the environment and creates ways of sustainable development. Social equability, this book series will enlighten the green technology in different ways, aspects and methods. This technology helps people to understand the use of different resources to fulfil needs and demands. Some points will be discussed as the combination of involuntary approaches, government incentives and a comprehensive regulatory framework will encourage the diffusion of green technology, least developed countries and developing small island states require unique support and measures to promote the green technologies.

Artificial Intelligence and Cybersecurity

Advances and Innovations

Edited by
Ishaani Priyadarshini
and Rohit Sharma

CRC Press
Taylor & Francis Group
Boca Raton London New York

CRC Press is an imprint of the
Taylor & Francis Group, an **informa** business

First edition published 2022
by CRC Press
6000 Broken Sound Parkway NW, Suite 300, Boca Raton, FL 33487-2742

and by CRC Press
2 Park Square, Milton Park, Abingdon, Oxon, OX14 4RN

Library of Congress Cataloging-in-Publication Data
Names: Priyadarshini, Ishaani, 1992- editor. | Sharma, Rohit, editor.
Title: Artificial intelligence and cybersecurity : advances and innovations
    / edited by Ishaani Priyadarshini and Rohit Sharma.
Description: First edition. | Boca Raton, FL : CRC Press, 2022. | Includes
    bibliographical references and index.
Identifiers: LCCN 2021038664 (print) | LCCN 2021038665 (ebook) | ISBN
    9780367466664 (hbk) | ISBN 9780367563950 (pbk) | ISBN 9781003097518
    (ebk)
Subjects: LCSH: Artificial intelligence--Industrial applications. |
    Artificial intelligence--Medical applications. | Machine
    learning--Industrial applications. | Blockchains (Databases) | Computer
    security.
Classification: LCC TA347.A78 A788 2022 (print) | LCC TA347.A78 (ebook) |
    DDC 006.3--dc23
LC record available at https://lccn.loc.gov/2021038664
LC ebook record available at https://lccn.loc.gov/2021038665

ISBN: 9780367466664 (hbk)
ISBN: 9780367563950 (pbk)
ISBN: 9781003097518 (ebk)

DOI: 10.1201/9781003097518

Typeset in Times
by Deanta Global Publishing Services, Chennai, India

Contents

Preface

This book will make the readers aware of the current state of the art of artificial intelligence in cybersecurity. It includes content related to various types of artificial intelligence techniques including machine learning and natural language processing and therefore highlights the impact of artificial intelligence in cybersecurity. Considering the current state, future research directions may be obtained from the book.

Chapter 1 discusses research into the classification of heart disease models from 303 datasets taken from the UCI repository. Before classification of data, cleaning and filtering was performed to eliminate the null values and identify the potential attributes. Among the four classification models: kNN, DT, XGBoost and random forest, the best classification models were identified on the basis of accuracy: area under ROC (AUC) and training time.

In **Chapter 2,** photonic crystals (PhC) have been utilised due to their generous structures and flexibility to adapt to every field. A dual core PCF refractive index sensor is proposed with the aim of improving performance, sensing ability, accuracy and selectivity by using identified AI algorithms. Tumor cells have refractive indices between 1.3342 and 1.4251. The maximum sensitivity observed was 32,358 nm/RIU and the minimum sensitivity observed was 11,258 nm/RIU. The highest accuracy reported for SVM was 96%.

The aim of **Chapter 3** is to provide a brief overview of security issues in the IoT and discuss possible countermeasures. This work highlights various security mechanisms adopted by IoT services and explores the most emerging IoT communication technology protocols. Further, this work provides the new scope to overcome threats and cyberattacks in an efficient and automated manner using computational intelligence technologies.

The goals of **Chapter 4** are to deliver an in-depth review of cybersecurity-oriented IoT architecture, provide insights into cyberthreats for the IoT, provide security necessities, outline the cyber-challenges and give visions on how these challenges can be overcome. Finally, the chapter offers new directions and research trends into cybersecurity and the IoT.

The focus of **Chapter 5** is to highlight the layered security architecture and working of the IoT. This work also throws light on the major threats and challenges faced by Industry 4.0 and other related technologies. Next, it outlines the proposed solutions to cyberthreats in Industry 4.0. Further, this work highlights the need for blockchain technology in Industry 4.0 and presents a detailed discussion on various application scenarios.

Chapter 6 discusses the political, social and economic issues plaguing farmers and causing huge economic losses. Research is focusing on detecting the problems in their early stages, thus saving farmers from financial hardship.

Chapter 7 provides a brief discussion on human emotion detection which reveals additional information. Several methods are discussed concerning the detection of human emotions from facial expressions. With the advancement in new technologies such as machine learning and deep learning, emotional features are extracted, showing the strength of both algorithms at pattern recognition and classification.

Chapter 8 elaborates on the significance of security systems and security measures employed in electronic commerce, discussing the essential requirements of e-transactions. Further, the chapter throws light on the various dimensions of security relating to e-commerce. Concepts such as secure SSL certificates, SHTTP and secure digital transmission as well as major strategies employed for combating fraud are discussed. The chapter wraps up with a review of e-commerce payment systems.

Chapter 9 reports on a survey that uses various techniques to detect, classify and quantify rice diseases. The authors stress the need for early detection of symptoms. Detection, classification and quantification are integrated into each technique's algorithms.

Chapter 10 proposes a transfer learning-based multimodal convolutional denoising auto encoder to perform multimodal compression and to reconstruct the data from its latent representation. Transfer learning helps the system to reuse the learned weights which may reconstruct the data with a better-quality score than by randomly initialised weights. The proposed work achieves a compression ratio of 128 and it is proved that multimodal compression is better than unimodal compression in cases of consuming multiple sensors. And the experimental result proves that the computation cost is lower in multimodal compression than in unimodal compression.

Chapter 11 discusses the popular contexts of metaheuristics and big data which are employed in current information technology. Later on, various key concepts of deep learning, deep neural networks and artificial neural networks are detailed. It critically analyses the difference between deep learning and machine learning while discussing the details of various metaheuristic algorithms such as genetic algorithms, practical swarm optimisation, etc.

Chapter 12 critically reviews communication strategies used on social networks from a multilingual perspective. This chapter first discusses communication strategies from different perspectives before making an in-depth review of communication strategies from a multilingual perspective. It then sheds light on the nature of communication on social networks as a multilingual and multicultural environment. Finally, it gives implications for educating second language communication in a multilingual world.

About the Editors

Ishaani Priyadarshini is a PhD candidate at the University of Delaware, USA. She earned her master's degree in cybersecurity from the University of Delaware. Prior to that she completed her bachelor's degree in computer science engineering and a master's degree in information security from Kalinga Institute of Industrial Technology, India. She has authored several book chapters for reputed publishers including IGI Global and Wiley & Sons and is also an author of several publications for SCIE indexed journals. As a certified reviewer, she conducts peer reviews of research papers for prestigious IEEE, Elsevier and Springer journals and is a member of the editorial board for the *International Journal of Information Security and Privacy* (IJISP). Her areas of research include cybersecurity, artificial intelligence and human–computer interaction.

Rohit Sharma is an assistant professor in the Department of Electronics and Communication Engineering, SRM Institute of Science and Technology, Delhi NCR Campus, Ghaziabad, India. He is an active member of ISTE, IEEE, ICS, IAENG and IACSIT. He is an editorial board member and reviewer of more than 12 international journals and conferences, including the premier journal *IEEE Access* and *IEEE Internet of Things Journal*. He serves as a book editor for seven different titles to be published by Apple Academic Press, CRC Press/Taylor & Francis Group, USA, Springer, etc. He received the Young Researcher Award at the 2nd Global Outreach Research and Education Summit & Awards 2019 hosted by Global Outreach Research & Education Association (GOREA). He serves as guest editor for the SCI journals of Elsevier, CEE and Springer WPC. He has been an organiser for various reputed international conferences. He served as an Editor and Organising Chair to the Third Springer International Conference on Microelectronics and Telecommunication (2019) and has served as the Editor and Organising Chair to the Second IEEE International Conference on Microelectronics and Telecommunication (2018); Editor and Organising Chair to IEEE International Conference on Microelectronics and Telecommunication (ICMETE–2016) held in India; Technical Committee member for CSMA2017, Wuhan, Hubei, China; EEWC 2017, Tianjin, China; IWMSE2017 Guangzhou, Guangdong, China; ICG2016, Guangzhou, Guangdong, China and ICCEIS2016 Dalian, Liaoning Province, China.

Contributors

Saurabh Bhatt
School of Engineering and Technology (SET)
Sharda University
Greater Noida, Uttar Pradesh, India

Bharat Bhushan
School of Engineering and Technology (SET)
Sharda University
Greater Noida, Uttar Pradesh, India

Pijush Dutta
Department of Electronics and Communication Engineering
Global Institute of Management and Technology
Krishnagar, India

Bui Phu Hung
University of Economics
Ho Chi Minh City, Vietnam

Agha Asim Husain
Department of ECE
I.T.S Engineering College
Greater Noida, India

Sushree Swagatika Jena
Computer Science & Information Technology
Siksha 'O' Anusandhan (Deemed to be University)
Bhubaneswar, India

Bui Thanh Khoa
Industrial University of Ho Chi Minh City
Ho Chi Minh City, Vietnam

M. Vinoth Kumar
Department of Electronics and Communication
SRM University

Tanmoy Maity
Department of MME
Indian Institute of Technology (ISM)
Dhanbad, India

Madhurima Majumder
Department of Electrical & Electronics Engineering
Mirmadan Mohanlal Government Polytechnic
Plassey, West Bengal, India

Ithayarani Pannerselvam
Department of Computer Science and Engineering
Koneru Lakshmaiah Education Foundation (KLEF) (Deemed to be University)
Vaddeswaram, Guntur District, India

Shobhandeb Paul
Guru Nanak Institute of Technology
Panihati, Kolkata, India

Sushree Bibhuprada B. Priyadarshini
Computer Science & Information Technology
Siksha 'O' Anusandhan (Deemed to be University)
Bhubaneswar, India

Mritunjay Rai
Department of MME
Indian Institute of Technology (ISM)
Dhanbad, India

S. Ramesh
Department of Electronics and Communication Engineering
Sri Shakthi Institute of Engineering and Technology
Coimbatore, India

Susmita Sen
Department of Electronics and
  Communication Engineering
Global Institute of Management and
  Technology
Krishnagar, India

Rohit Sharma
SRM Institute of Science and
  Technology
Ghaziabad, India

Sunil Sharma
Department of Electronics Engineering
Rajasthan Technical University
Kota, India

Neha Shaw
Department of Electronics and
  Communication Engineering
Global Institute of Management and
  Technology
Krishnagar, India
and
Department of Electrical & Electronics
  Engineering

**Mirmadan Mohanlal Government
  Polytechnic**
Plassey, India

Tanya Srivastava
School of Engineering and Technology
Sharda University
Greater Noida, Uttar Pradesh, India

Lokesh Tharani
Department of Electronics Engineering
Rajasthan Technical University
Kota, India

Gagan Varshney
School of Engineering and Technology
Sharda University
Greater Noida, Uttar Pradesh, India

R. K. Yadav
Department of Electronics &
  Communication Engineering
RKGIT
Ghaziabad, India

1 Heart Disease Prediction
A Comparative Study Based on a Machine-Learning Approach

Pijush Dutta, Shobhandeb Paul, Neha Shaw, Susmita Sen and Madhurima Majumder

CONTENTS

INTRODUCTION

According to the WHO chronic disease is one of the common causes of death for men and women across the globe. The most common reasons for heart disease are the use of tobacco and alcohol, an unhealthy diet or lack of physical exercise. These cause an increase in blood pressure, body mass index, etc. As a result, there will be a high chance of heart disease, such as heart attacks, strokes and heart failure-like complications. In real life to predict whether a person has heart disease or not a consultant takes a long time to perform several tests, even if sometimes they predict incorrectly due to insufficient skilled knowledge to interpret the data. The diagnosis of coronary illness and its treatment are intricate because of the poor accessibility of diagnostic apparatus and a deficiency of doctors and other assets that influence the appropriate diagnosis and treatment of heart patients (Ghwanmeh et al., 2013). A precise and legitimate heart diagnosis can reduce the related dangers of extreme heart issues in a patient (Al-Shayea, 2011).

According to a WHO survey, almost 80% of cardiovascular disease (CVD) deaths take place in underdeveloped and developing middle-income countries. Consequently, there is an extraordinary need to detect the disease at a timely stage to combat this disturbing issue. One-third of the world's population suffers from heart disease. Age, sex, smoking, heredity risk factors (hypertension, diabetes), cholesterol, body weight, etc. are all symptoms of coronary illness. Among all this, diet, physical weight and physical dimensions are viewed as dangerous factors (Gudadhe et al., 2010) which can be controllable. It is difficult to take a physical stand against the chances of getting coronary illness dependent on hazard factors (Amin et al., 2013; Muntner et al., 2005).

For the most part, data mining assumes a fundamental function in the forecast of sickness in the medical services industry. Sophisticated data analysis equipment is needed to extract valuable information from an enormous amount of clinical information. Clinical information is an ever-developing wellspring of data created from patient records at emergency clinics. In any case, the data covered in these records is a gigantic asset bank for clinical investigation. In our work, the point is to take a medical choice which is an exceptionally specific testing position because of different components, particularly on account of illnesses that show comparative indications or concern uncommon illnesses. It is a significant focus of artificial intelligence (AI) in medicine. An AI framework would take the patient's information and propose suitable tests. The framework can remove confidential information from a verifiable clinical information base, can anticipate patients' ailments and utilise the clinical profiles, for example, age, blood pressure, blood sugar and so on; it can predict the probability of patients developing a disease. It is found that with access to a huge amount of patient data that contain valuable knowledge of medical issues, it is very useful to use AI in the diagnosis of disease. Motivated by this, there is much research conducted on AI, machine learning and data mining to diagnose diseases. Some of them are: Parkinson's disorder of the central nervous system (Khan et al., 2018; Sadek et al., 2019),, prediction of diabetes (El_Jerjawi & Abu-Naser, 2018; Gadekallu & Khare, 2017), SARS COVID-19 (Lalmuanawma et al., 2020), diagnosis of lung diseases (Das et al., 2018), breast cancer (Khan et al., 2018), heart disease (Maji & Arora, 2019), liver disease (Nahar & Ara, 2018), diabetic kidney disease (Makino et al., 2019), prediction of disease using simulated annealing back propagations (Hu et al., 2018), special disease (diabetes and cancer) prediction (Kanchan & Kishor, 2016), precision of cardiology (Johnson et al., 2018), prediction of ocular disease (Schmidt-Erfurth et al., 2018), prediction of tumor category (Nasser & Abu-Naser, 2019), blood donation prediction (Alajrami et al., 2019), diagnosis of tuberculosis (Dande & Samant, 2018), lung cancer prediction (Chauhan & Jaiswal, 2016), diagnosis of coronary artery disease, prediction of thyroid disease (Ioniță & Ioniță, 2016), mental illness (Graham et al., 2019).

Inspired by the above, in this work we propose supervised machine-learning algorithms for modelling and predicting heart disease. As the huge number of patient datasets obtained from healthcare professionals has a specific output, that is either the patient has heart disease or does not, corresponding to 13 potential input attributes such as age, sex, chest pain, resting blood pressure, fasting blood sugar, electrocardiographic result, maximum heart rate, cholesterol, exercise-induced angina and diagnosis of any heart disease, so here we applied the classification algorithm to predict whether the person has heart disease or not using four different classification algorithms: k-nearest neighbor, decision tree, XGBoost and random forest.

The remaining part of the chapter is structured as follows: the first sections are the introduction followed by a literature survey where previous work and algorithms are explained in tabular form. The section 'Materials and Methods' describes the different attributes of the datasets, data processing and feature selections. In the methodology section, all the four supervised machine-learning techniques which include the application, advantages and disadvantages of each algorithm, are covered. In the section 'Performance Analysis', we describe the confusion matrix of all the algorithms which includes five performance indexes: accuracy, specificity, sensitivity, F1 score and execution time. Finally, this results in an analysis followed by a conclusion.

LITERATURE SURVEY

In this section we highlight some state-of-the-art models and their results in the field of cardiovascular disease. Meanwhile, different investigations give just a brief look into anticipating coronary illness utilising AI procedures. This section investigates the methods that are identified with the proposed approach (Table 1.1).

AI DIAGNOSIS PROCESS

In this AI model we used four suitable machine-learning (ML) strategies that are normally selected and upgraded relying on the application to diagnose the heart condition (yield) in view of the indicators. Additional indicators (for example, test datasets) are incorporated into the ML model to confirm whether or not the individual has any coronary illness and improve the model performance shown in Figure 1.1.

DATASETS AND ATTRIBUTES

To perform this research the chronic disease datasets were obtained from the UCI Repository of Machine Learning Databases which contains a total of 303 records with 12 potential attributes and two attributes for the patient's personal information (Abdeldjouad et al., 2020; Palechor et al., 2017).

DATA CLEANING

Data cleaning is performed on the datasets, as six out of 3030 rows contain null values which reduce the accuracy of the predictive model.

TABLE 1.1
Notable ML Methods for Cardiovascular Disease

Sl No	Year Publishing	Paper	Algorithm Used	Disease	Outcomes
1	June, 2019	(Mohan et al., 2019)	Hybrid random forest with a linear model (HRFLM)	Cardiovascular diseases (CVD)	This model accurate upto 88.70%
2	July, 2017	(Alić et al., 2017)	Artificial neural network (ANNs) and naïve Bayesian (NB) network	Diabetes and cardiovascular diseases (CVD)	Highest accuracy offered by naïve Bayesian network for classification of diabetes and CVD, 99.51% and 97.92% retrospectively
3	May, 2020	(Muniasamy et al., 2020)	Artificial neural network (ANN), support vector machine (SVM), decision tree (DT), linear discriminant analysis (LDA), nearest neighbor (NN) and random forest (RF)	Heart diseases	RF, LDA, DT and ANN outperformed SVM and NN by means of accuracy, F1 score, precision and error rate
4	July, 2020	(Pollard et al., 2020)	Random forests, convolutional neural network, and lasso models	Detection of cardiovascular disease using ECG	CVD detection is feasible
5	August, 2019	(Mezzatesta et al., 2019)	Support vector classifier (SVC)with radial basic function (RBF) kernel algorithm	Cardiovascular diseases in dialysis patients	Avow the factors to grant an accuracy of 95.25% in the Italian dataset and of 92.15% in the American dataset
6	July, 2020	(Nayan et al., 2020)	linear discriminant analysis (LDA), SVM, DT KNN and ANN	Cardiovascular disease prediction from ECG	Among this five ML ANNs achieved the maximum accuracy of 90% to predict the model
7	March, 2017	(Bhatt et al., 2017)	J48 and naïve Bayes algorithm	Analyse cardiovascular disease	The enforcement of both the algorithms affirms the veracity of the algorithm and anticipated the various causes of heart disease

(Continued)

TABLE 1.1 (CONTINUED)
Notable ML Methods for Cardiovascular Disease

Si No	Year Publishing	Paper	Algorithm Used	Disease	Outcomes
8	January, 2020	(Martin-Isla et al., 2020)	Logistic regression (LR), support vector machine (SVM), random forest (RF), artificial neural network (ANN) and convolutional neural network (CNN)	Image-based cardiac diagnosis	Review of mentioned algorithm on cardiac diagnosis based on images
9	February, 2017	(Palechor et al., 2017)	Bayesian networks (BN), KNN, DT and SVM	Cardiovascular conditions	SVM outperformed by means of precision
10	January, 2019	(Li et al., 2019)	Adaboost + RF	Cardiovascular disease	Four key indicators of the Adaboost + RF model are better than other machine learning algorithms for unbalanced missing datasets
11	October, 2019	(Maiga & Hungilo, 2019)	RF, NB, KNN and logistic regression (LR)	Cardiovascular diseases (CVDs) are ailments of heart and blood vessels	RF actualises high categorisation of accuracy of 73 %.
12	2020	(Ramkumar et al., 2020)	Hybrid random forest with linear model (HRFLM)	Cardiovascular disease	This will come through for the embellished performance level with a higher acuteness level about 92%

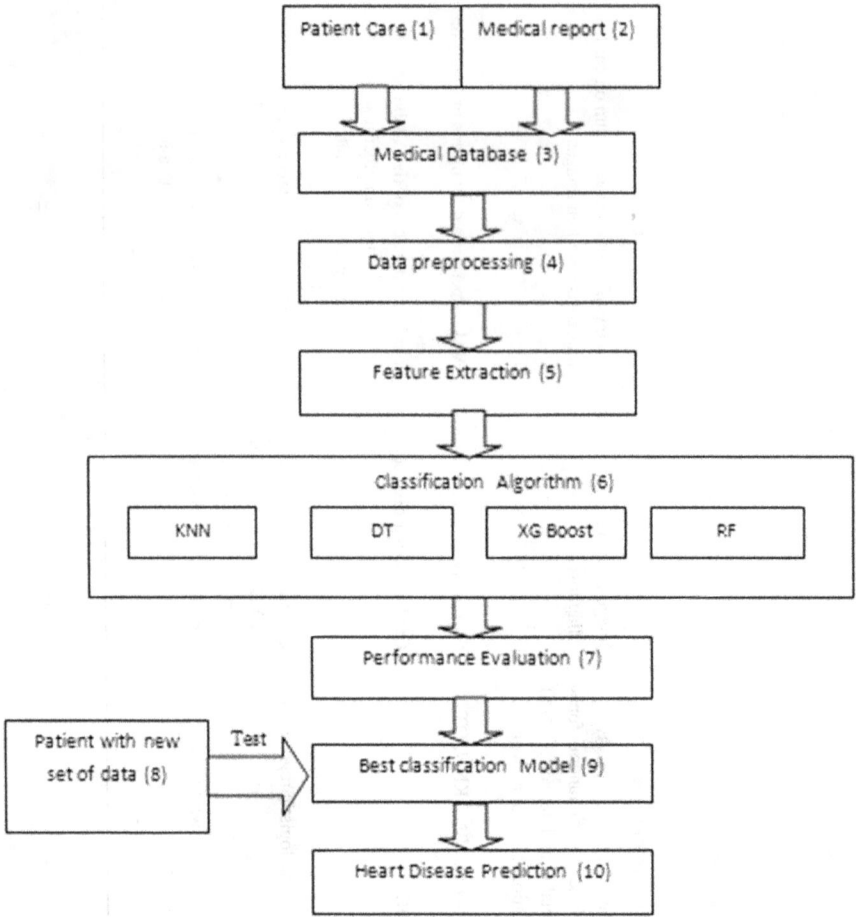

FIGURE 1.1 Flowchart of the proposed model.

FEATURE SELECTION

In this process the relevant features are extracted from original datasets. Among three different techniques here we use the filtering method. Among 14 attributes two attributes are removed as they convey the personal information (sex and age) of the patient. Figure 1.2 shows the features after the data cleaning.

MACHINE LEARNING

ML alludes to the utilisation of a computational breakthrough that can figure out how to perform given functions from model information without the need for directions. This aspect of AI utilises advanced statistical methods to remove prescient

	age	sex	cp	trestbps	chol	fbs	restecg	thalach	exang	oldpeak	slope	ca	thal	target
0	63	1	3	145	233	1	0	150	0	2.3	0	0	1	1
1	37	1	2	130	250	0	1	187	0	3.5	0	0	2	1
2	41	0	1	130	204	0	0	172	0	1.4	2	0	2	1
3	56	1	1	120	236	0	1	178	0	0.8	2	0	2	1
4	57	0	0	120	354	0	1	163	1	0.6	2	0	2	1

FIGURE 1.2 Feature selection of the datasets.

	age	trestbps	chol	thalach	oldpeak	target	sex_0	sex_1	cp_0	cp_1	...	slope_2	ca_0	ca_1	ca_2	ca_3	ca_4	thal_0	thal_1
0	0.952197	0.763956	-0.256334	0.015443	1.087338	1	0	1	0	0	...	0	1	0	0	0	0	0	1
1	-1.915313	-0.092738	0.072199	1.633471	2.122573	1	0	1	0	0	...	0	1	0	0	0	0	0	0
2	-1.474158	-0.092738	-0.816773	0.977514	0.310912	1	1	0	0	1	...	1	1	0	0	0	0	0	0
3	0.180175	-0.663867	-0.198357	1.239897	-0.206705	1	0	1	0	1	...	1	1	0	0	0	0	0	0
4	0.290464	-0.663867	2.082050	0.583939	-0.379244	1	1	0	1	0	...	1	1	0	0	0	0	0	0

FIGURE1.3 Datasets after removing the dummy variables and data cleaning.

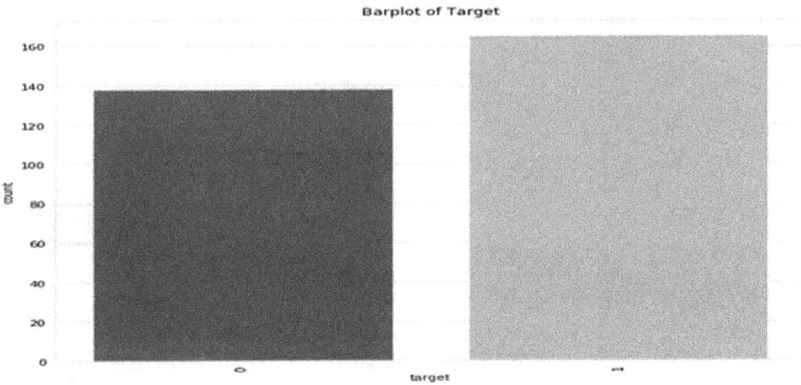

FIGURE 1.4 Plot for the target value.

or unfair examples from the preparation information to play out the most precise forecasts on new information. In this research using the python platform, distinctive prescient calculations were picked to assemble the model, namely: k-nearest neighbor (kNN), random forest (RF), decision tree (DT) and XGBoost.

In order to maintain uniformity, some features are given dummy variables. A view of the datasets after elimination of those dummy variables and data cleaning is shown in Figure 1.3 while Figures 1.4 and 1.5 show the bar plot of the target value and histogram graphs for the related features of the datasets.

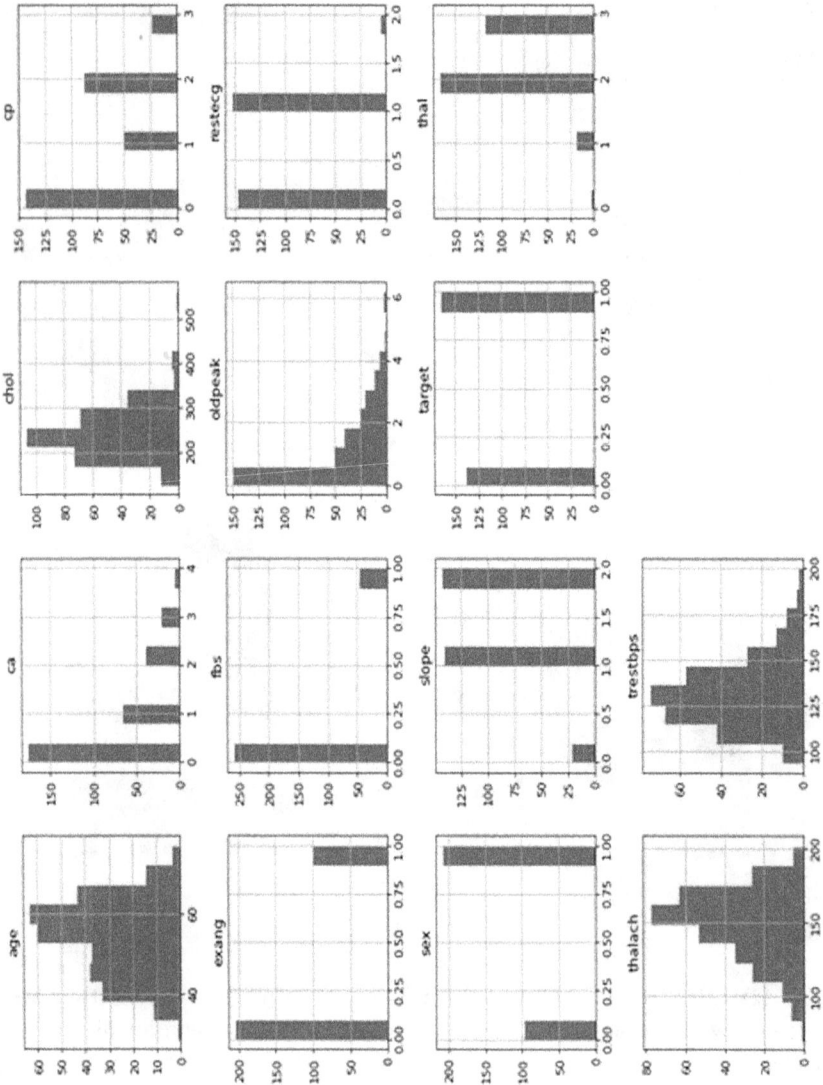

FIGURE 1.5 Histogram of the datasets.

XGBOOST

XGBoost is an improved algorithm compared with a decision tree, which can increase the number of trees with equal processing time (Chen et al., 2018).It has the following advantages: (1) it uses the second-order Taylor expression to estimate the purpose of the function, making it simpler to locate the ideal arrangement; (2) it can deal with scant and missing data; (3) it creates a choice tree utilising the basic score; (4) it reduces the processing time and increases the number of nodes. The fundamental hindrance is when the volume of information is large, the strategy is tedious (Chen et al., 2018; Rahman et al., 2020). This improved algorithm is applied in different fields of research; some of them are crude oil pricing (Gumus & Kiran, 2017), image classification (Ren et al., 2019), and disease diagnosis (Ogunleye & Qing-Guo, 2019).

RESULT ANALYSIS

The dataset contained 303 datasets taken from the UCI Machine Learning Repository, through which we need to train the model using suitable algorithms. After analysing the datasets, it was found that the dataset is a type of classification problem, thus the main classification algorithms were applied to train the model and obtain accuracy. To play out all these three algorithms we utilised the following features – processor and platform: Intel i3, sixth era processor, OS: Ubuntu 20.04 and RAM 8 GB, python 3.7.6, and Jupyter journal 6.03. Three prescient models were created utilising logistic regression, decision tree, and random forest separately. When the ML exemplar was prepared and tested, various metrics were acquired to assess its utility.

CONFUSION MATRIX

Accuracy measures the level of the calculation, arranging the information accurately. It is a basic measure utilised in numerous logical situations if there is no class awkwardness. One of the disadvantages of utilising precision as the measurement is that there is an information error when estimating improper positive and untruthful negative perceptions. Accordingly, exactitudes and susceptibility are generally utilised for estimating the presentation of the algorithm for potential class unevenness. To survey the exhibition of a calculation and to understand it, a tabular report called a confusion matrix is implied (Dutta et al., 2021; Luque et al., 2019). Figures 1.6–1.9 show the classification reports of classification models utilised in heart disease prediction. Figure 1.10 shows the comparative study of the four ML techniques on the basis of confusion matrix performance parameters.

kNN classification model performance depends upon the value of neighbor number k conventionally as the number k increases the boundary level of the classification, becoming smoother up to a certain level. From Figure 1.11 it is seen that as the value increases the classification score also increases and it reaches maximum when $k = 9$, afterwards the score value decreases for the dataset characteristics (Figure 1.12).

```
Classification Report
                precision    recall  f1-score   support

           0         0.85      0.83      0.84        41
           1         0.86      0.88      0.87        50

    accuracy                            0.86        91
   macro avg         0.86      0.85      0.86        91
weighted avg         0.86      0.86      0.86        91

accuracy:

0.8571428571428571
```

FIGURE 1.6   Classification report of kNN algorithm.

```
Classification Report
                precision    recall  f1-score   support

           0         0.73      0.78      0.75        41
           1         0.81      0.76      0.78        50

    accuracy                            0.77        91
   macro avg         0.77      0.77      0.77        91
weighted avg         0.77      0.77      0.77        91

accuracy:

0.7692307692307693
```

FIGURE 1.7   Classification report of decision tree.

RECEIVER OPERATING CURVE (ROC)

A receiver operating curve (ROC) is one type of characteristic plot in ML which extricates the properties from the confusion matrix, specificity and sensitivity (Koen et al., 2017; Mas, 2018; Safari et al., 2016). It is plotted against a true-positive rate (sensitivity, recall or probability of correctness) with a true negative (error probability). Area under ROC (AUC) is mainly used in ML to quantify the algorithm enforcement. It is observable that AUC can be obtained from choice limits from ML models in spite of the way that it is prepared with diverse yields. At the point when a prepared model is approached to make a forecast, likelihood can be figured and used to create a ROC plot. Figure 1.6 shows the comparative study of four ML algorithms on the basis of AUC, where it is seen that random forest occupied the maximum AUC of 0.95 followed by kNN of 0.93 also shown in Table 1.2.

Figure 1.13 shows the graphical comparison of the training and testing computational time for all the algorithms mentioned and used in this research. It is seen

```
Classification Report
               precision     recall   f1-score    support

          0       0.80       0.80       0.80         41
          1       0.84       0.84       0.84         50

   accuracy                             0.82         91
  macro avg       0.82       0.82       0.82         91
weighted avg      0.82       0.82       0.82         91

accuracy:

0.8241758241758241
```

FIGURE 1.8   Classification report of XGBoost.

```
Classification Report
               precision     recall   f1-score    support

          0       0.79       0.83       0.81         41
          1       0.85       0.82       0.84         50

   accuracy                             0.82         91
  macro avg       0.82       0.82       0.82         91
weighted avg      0.83       0.82       0.82         91

accuracy:

0.8241758241758241
```

FIGURE 1.9   Classification report of random forest.

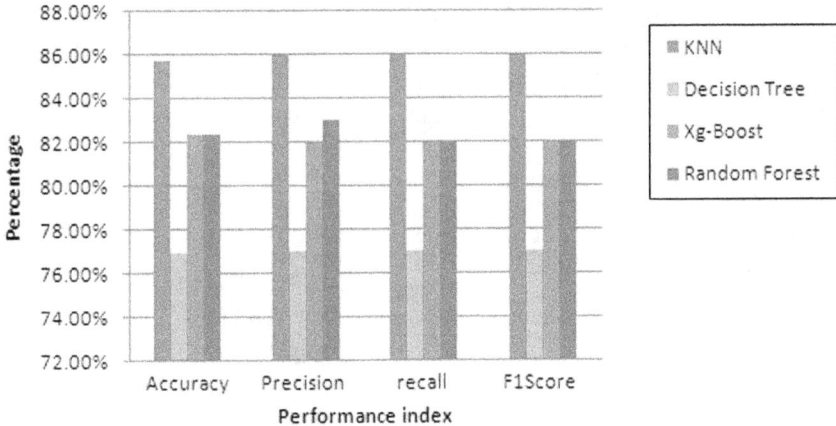

FIGURE 1.10   Comparative study based on performance index.

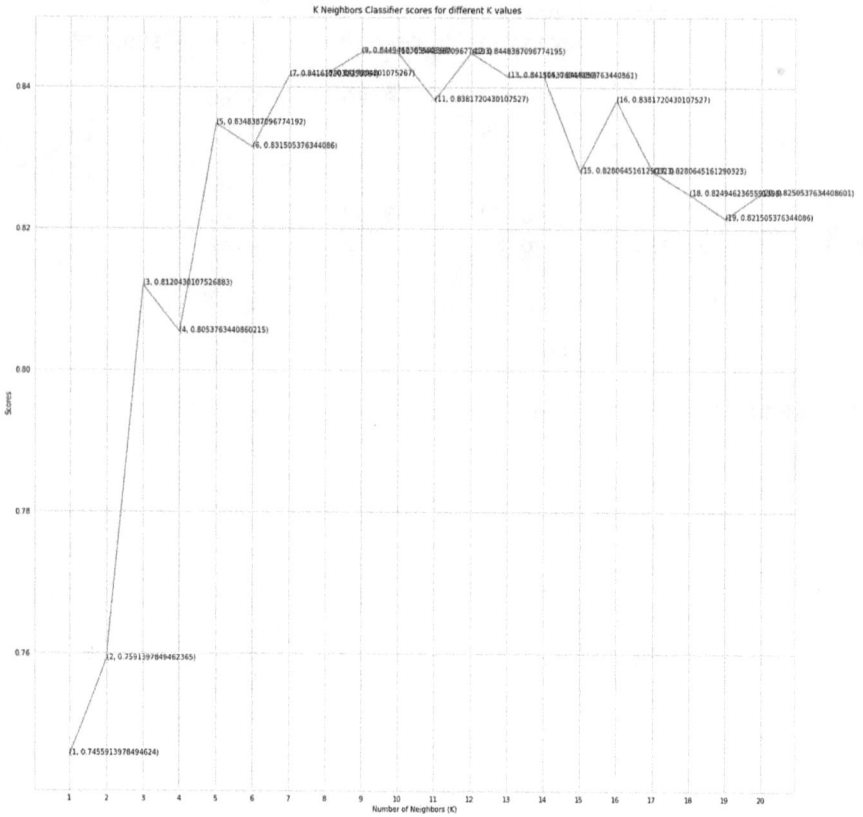

FIGURE 1.11   K neighbors classifier scores for different K values.

that kNN has the least training time while XGBoost has the least testing time. Figure 1.7 shows the graphical comparison of four performance indexes of the confusion matrix. All the performance indexes including certainty, particularity, retrospective idea andF1 score are highest for kNN followed by random forest. Table 1.3 indicates the entire performance index offered by kNN, DT, XGBoost and RF.

CONCLUSION AND FUTURE SCOPE

At present, AI performs a significant role in the productive arrangement of the healthcare dataset. Proper diagnosis and prediction of chronic diseases can be improved if the characteristics, nature of the data and potential attributes are clearly identified from the medical dataset. To prepare the model, datasets are taken from the UCI Machine Learning Repository. To avoid the problem of overfitting and to reduce the computational time in this research we used four distinctive classification algorithms like k-nearest neighbor, decision tree, XGBoost and random forest.

ROC curve

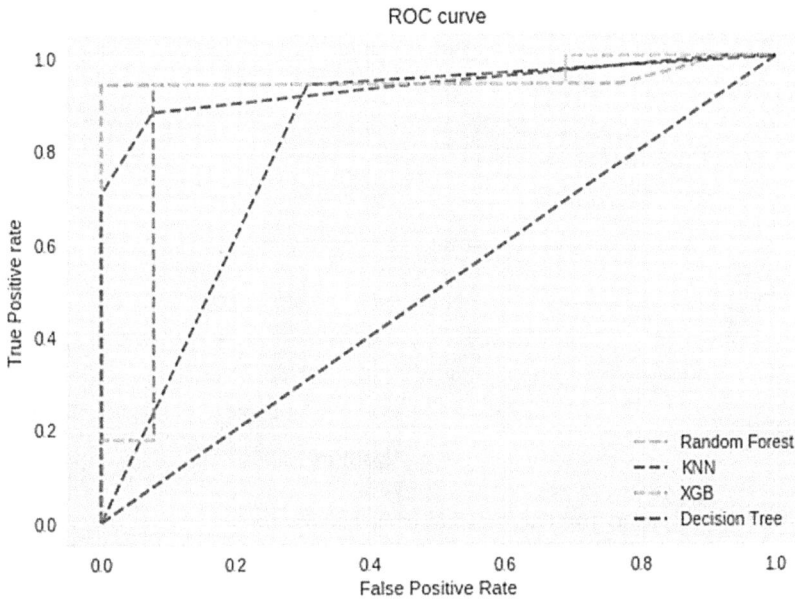

FIGURE 1.12 Comparative study based on ROC.

TABLE 1.2
Comparative Study Based on Overall Performance

Algorithm	Training Time	Accuracy	AUC
kNN	0.007	85.71%	0.93
Decision tree	0.011	76.92%	0.8167
XGBoost	0.331	82.41%	0.9004
Random forest	0.461	82.41%	0.95

Accuracy, sensitivity, specificity and training and testing execution time are used as performance metrics for the assessment of these algorithms.

The performance index measures using in the classical algorithm include accuracy, precision, recall, F1 score, training time and area under ROC (AUC). From Table 1.2 it is seen that training time and accuracy are better for kNN than RF, XGBoost and DT, while AUC of RF is better followed by the kNN algorithm. The overall result analysis concludes two main points: (1) machine-learning kNN can classify and diagnose cardiovascular disease (CVD) patient data very accurately and (2) we compared the performance of kNN with that of DT, XGBoost and RF and observed that kNN outperforms DT, XGBoost and RF for this particular application example. This method is applicable for any type of medical data diagnostics where a given set of appropriate input attributes classifies the person having or not having any disease. In other words, to apply this method the learning component needs to

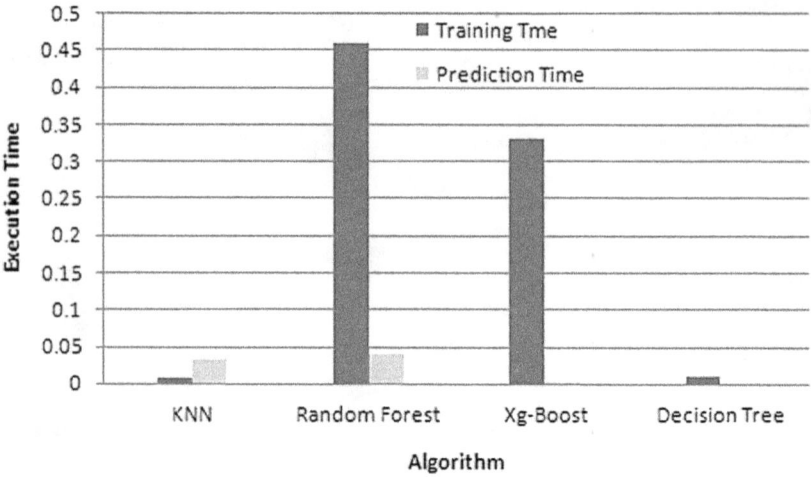

FIGURE 1.13 Comparative study based on computational time.

TABLE 1.3

Comparative Study on Performance Index Based on Confusion Matrix

Algorithm	Accuracy	Precision	Recall	F1 Score
kNN	85.71%	86%	86%	86%
Decision tree	76.92%	77%	77%	77%
XGBoost	82.41%	82%	82%	82%
Random forest	82.41%	83%	82%	82%

be trained on the behavior of the combination of the selected input attributes so the model can easily classify the patient characteristics.

Although a few endeavors and continuous research exist on conveying knowledge into a calculation's dynamic, (for example, heatmaps), these endeavors are not adequately expounded so far as to persuade most cardiology professionals to utilise it as an asymptomatic black box in everyday administration. The future work can be described as follows. (1) Analysing the exhibition (accuracy) of the framework using parameter tuning. (2) Implementing an ensemble-based classifier to improve the exhibition of the framework.(3) Using a hybrid IoT-based ML classifier to improve the model. (4) Using oversampling reduction techniques to improve performance of the proposed model.

ACKNOWLEDGEMENTS

We recognise the great contribution of the UCI Machine Learning Repository for providing datasets of information.

BIBLIOGRAPHY

Abdeldjouad, F. Z., Brahami, M. & Matta, N. (2020). A hybrid approach for heart disease diagnosis and prediction using machine learning techniques. In M. Jmaiel, M. Mokhtari, B. Abdulrazak, H. Aloulou & S. Kallel (Eds), *The impact of digital technologies on public health in developed and developing countries* (Vol. 12157, pp. 299–306). Springer International Publishing. https://doi.org/10.1007/978-3-030-51517-1_26

Alajrami, E., Abu-Nasser, B. S., Khalil, A. J., Musleh, M. M., Barhoom, A. M. & Naser, S. A. (2019). Blood donation prediction using artificial neural network. *International Journal of Academic Engineering Research*, *3*(10), 1–7.

Alić, B., Gurbeta, L. & Badnjević, A. (2017). Machine learning techniques for classification of diabetes and cardiovascular diseases. In *2017 6th Mediterranean Conference on Embedded Computing (MECO)*, 1–4.

Al-Shayea, Q. K. (2011). Artificial neural networks in medical diagnosis. *International Journal of Computer Science Issues*, *8*(2), 150–154.

Amin, S. U., Agarwal, K. & Beg, R. (2013). Genetic neural network based data mining in prediction of heart disease using risk factors. In *2013 IEEE Conference on Information & Communication Technologies*, 1227–1231.

Azad, C., Bhushan, B., Sharma, R. et al. (2021). Prediction model using SMOTE, genetic algorithm and decision tree (PMSGD) for classification of diabetes mellitus. *Multimedia Systems*. https://doi.org/10.1007/s00530-021-00817-2

Begam, S., Vimala, J., Selvachandran, G., Ngan, T. T. & Sharma, R. (2020). Similarity measure of lattice ordered multi-fuzzy soft sets based on set theoretic approach and its application in decision making. *Mathematics*, *8*, 1255.

Bhatt, A., Dubey, S. K., Bhatt, A. K. & Joshi, M. (2017). Data mining approach to predict and analyze the cardiovascular disease. In *Proceedings of the 5th International Conference on Frontiers in Intelligent Computing: Theory and Applications*, 117–126.

Chauhan, D. & Jaiswal, V. (2016). An efficient data mining classification approach for detecting lung cancer disease. In *2016 International Conference on Communication and Electronics Systems (ICCES)*, 1–8.

Chen, Z., Jiang, F., Cheng, Y., Gu, X., Liu, W. & Peng, J. (2018). XGBoost classifier for DDoS attack detection and analysis in SDN-based cloud. In *2018 IEEE International Conference on Big Data and Smart Computing (BigComp)*. https://doi.org/10.1109/BigComp.2018.00044

Dande, P. & Samant, P. (2018). Acquaintance to artificial neural networks and use of artificial intelligence as a diagnostic tool for tuberculosis: A review. *Tuberculosis*, *108*, 1–9.

Dansana, D., Kumar, R., Parida, A., Sharma, R., Adhikari, J. D. et al. (2021). Using susceptible-exposed-infectious-recovered model to forecast coronavirus outbreak. *Computers, Materials & Continua*, *67*(2), 1595–1612.

Das, N., Topalovic, M. & Janssens, W. (2018). Artificial intelligence in diagnosis of obstructive lung disease: Current status and future potential. *Current Opinion in Pulmonary Medicine*, *24*(2), 117–123.

Dutta, P., Paul, S. & Kumar, A. (2021). Comparative analysis of various supervised machine learning techniques for diagnosis of COVID-19. In *Electronic devices, circuits, and systems for biomedical applications* (pp. 521–540). https://doi.org/10.1016/B978-0-323-85172-5.00020-4

El_Jerjawi, N. S. & Abu-Naser, S. S. (2018). Diabetes prediction using artificial neural network. *International Journal of Advance Science and Technology*, *121*, 55–64.

Gadekallu, T. R. & Khare, N. (2017). Cuckoo search optimized reduction and fuzzy logic classifier for heart disease and diabetes prediction. *International Journal of Fuzzy System Applications (IJFSA)*, *6*(2), 25–42.

Ghanem, S., Kanungo, P., Panda, G. et al. (2021). Lane detection under artificial colored light in tunnels and on highways: an IoT-based framework for smart city infrastructure. *Complex & Intelligent Systems.* https://doi.org/10.1007/s40747-021-00381-2

Ghwanmeh, S., Mohammad, A. H. & Alibrahim, A. (2013). Innovative artificial neural networks-based decision support system for heart diseases diagnosis. *Adel Hamdan Abu Arrah—Academia.edu.* https://www.academia.edu/18693647/Innovative_Artificial_Neural_Networks_Based_Decision_Support_System_for_Heart_Diseases_Diagnosis

Graham, S., Depp, C., Lee, E. E., Nebeker, C., Tu, X., Kim, H.-C. & Jeste, D. V. (2019). Artificial intelligence for mental health and mental illnesses: An overview. *Current Psychiatry Reports, 21*(11), 116.

Gudadhe, M., Wankhade, K. & Dongre, S. (2010). Decision support system for heart disease based on support vector machine and artificial neural network. In *2010 International Conference on Computer and Communication Technology (ICCCT)*, 741–745.

Gumus, M. & Kiran, M. S. (2017). Crude oil price forecasting using XGBoost. In *2017 International Conference on Computer Science and Engineering (UBMK)*, 1100–1103.

Hu, F., Wang, M., Zhu, Y., Liu, J. & Jia, Y. (2018). A time simulated annealing-back propagation algorithm and its application in disease prediction. *Modern Physics Letters B, 32*(25), 1850303.

Ioniţă, I. & Ioniţă, L. (2016). Prediction of thyroid disease using data mining techniques. *BRAIN. Broad Research in Artificial Intelligence and Neuroscience, 7*(3), 115–124.

Jha, S., et al. (2019). Deep learning approach for software maintainability metrics prediction. *IEEE Access, 7*, 61840–61855.

Johnson, K. W., Soto, J. T., Glicksberg, B. S., Shameer, K., Miotto, R., Ali, M., Ashley, E. & Dudley, J. T. (2018). Artificial intelligence in cardiology. *Journal of the American College of Cardiology, 71*(23), 2668–2679.

Kanchan, B. D. & Kishor, M. M. (2016). Study of machine learning algorithms for special disease prediction using principal of component analysis. In *2016 International Conference on Global Trends in Signal Processing, Information Computing and Communication (ICGTSPICC)*, 5–10.

Khan, M. M., Mendes, A. & Chalup, S. K. (2018). Evolutionary wavelet neural network ensembles for breast cancer and Parkinson's disease prediction. *PLoS One, 13*(2), e0192192.

Koen, J. D., Barrett, F. S., Harlow, I. M. & Yonelinas, A. P. (2017). The ROC toolbox: A toolbox for analyzing receiver-operating characteristics derived from confidence ratings. *Behavior Research Methods, 49*(4), 1399–1406.

Lalmuanawma, S., Hussain, J. & Chhakchhuak, L. (2020). Applications of machine learning and artificial intelligence for Covid-19 (SARS-CoV-2) pandemic: A review. *Chaos, Solitons & Fractals, 139*, 110059.

Li, R., Shen, S., Chen, G., Xie, T., Ji, S., Zhou, B. & Wang, Z. (2019). Multilevel risk prediction of cardiovascular disease based on adaboost+ RF ensemble learning. *IOP Conference Series: Materials Science and Engineering, 533*(1), 012050.

Luque, A., Carrasco, A., Martín, A. & de las Heras, A. (2019). The impact of class imbalance in classification performance metrics based on the binary confusion matrix. *Pattern Recognition, 91*, 216–231.

Maiga, J. & Hungilo, G. G. (2019). Comparison of machine learning models in prediction of cardiovascular disease using health record data. In *2019 International Conference on Informatics, Multimedia, Cyber and Information System (ICIMCIS)*, 45–48.

Maji, S. & Arora, S. (2019). Decision tree algorithms for prediction of heart disease. In Simon Fong, Shyam Akashe & Parikshit N. Mahalle (eds.) *Information and communication technology for competitive strategies* (pp. 447–454). Springer.

Makino, M., Yoshimoto, R., Ono, M., Itoko, T., Katsuki, T., Koseki, A., Kudo, M., Haida, K., Kuroda, J. & Yanagiya, R. (2019). Artificial intelligence predicts the progression of diabetic kidney disease using big data machine learning. *Scientific Reports, 9*(1), 1–9.

Malik, P., et al. (2021). Industrial internet of things and its applications in industry 4.0: State of the art. *Computer Communication*, Elsevier, *166*, 125–139.

Martin-Isla, C., Campello, V. M., Izquierdo, C., Raisi-Estabragh, Z., Baeßler, B., Petersen, S. E. & Lekadir, K. (2020). Image-based cardiac diagnosis with machine learning: A review. *Frontiers in Cardiovascular Medicine, 7,* 1.

Mas, J. F. (2018). Receiver operating characteristic (roc) analysis. In María Teresa Camacho Olmedo, Martin Paegelow, Jean-François Mas & Francisco Escobar (eds.) *Geomatic approaches for modeling land change scenarios* (pp. 465–467). Springer.

Mezzatesta, S., Torino, C., De Meo, P., Fiumara, G. & Vilasi, A. (2019). A machine learning-based approach for predicting the outbreak of cardiovascular diseases in patients on dialysis. *Computer Methods and Programs in Biomedicine, 177,* 9–15.

Mohan, S., Thirumalai, C. & Srivastava, G. (2019). Effective heart disease prediction using hybrid machine learning techniques. *IEEE Access, 7,* 81542–81554.

Muniasamy, A., Muniasamy, V. & Bhatnagar, R. (2020). Predictive analytics for cardiovascular disease diagnosis using machine learning techniques. In *International Conference on Advanced Machine Learning Technologies and Applications*, 3rd February 2020, Singapore, 493–502.

Muntner, P., He, J., Astor, B. C., Folsom, A. R. & Coresh, J. (2005). Traditional and nontraditional risk factors predict coronary heart disease in chronic kidney disease: Results from the atherosclerosis risk in communities study. *Journal of the American Society of Nephrology, 16*(2), 529–538.

Nahar, N. & Ara, F. (2018). Liver disease prediction by using different decision tree techniques. *International Journal of Data Mining & Knowledge Management Process, 8*(2), 01–09.

Nasser, I. M., & Abu-Naser, S. S. (2019). Predicting tumor category using artificial neural networks. *International Journal of academic Health and Health Research, 3*(2), 1–7.

Nayan, N. A., Hamid, H. A., Suboh, M. Z., Jaafar, R., Abdullah, N., Yusof, N. A. M., Hamid, M. A., Zubiri, N. F., Arifin, A. S. K. & Daud, S. M. A. (2020). Cardiovascular disease prediction from electrocardiogram by using machine learning. *International Journal of Online & Biomedical Engineering, 16*(7), 1–12.

Nguyen, P. T., Ha, D. H., Avand, M., Jaafari, A., Nguyen, H. D., Al-Ansari, N., Van Phong, T., Sharma, R., Kumar, R., Le, H. V., Ho, L. S., Prakash, I. & Pham, B. T. (2020). Soft computing ensemble models based on logistic regression for groundwater potential mapping. *Applied Sciences,10,* 2469.

Ogunleye, A. A. & Qing-Guo, W. (2019). XGBoost model for chronic kidney disease diagnosis. *IEEE/ACM Transactions on Computational Biology and Bioinformatics, 17*(6), 2131–2140.

Palechor, F. M., De la Hoz Manotas, A., Colpas, P. A., Ojeda, J. S., Ortega, R. M. & Melo, M. P. (2017). Cardiovascular disease analysis using supervised and unsupervised data mining techniques. *JSW, 12*(2), 81–90.

Pollard, J., Haq, K. T., Lutz, K., Rogovoy, N., Paternostro, K., Soliman, E., Maher, J., Lima, J., Musani, S. & Tereshchenko, L. G. (2020). Using ECG machine learning for detection of cardiovascular disease in African American men and women: The Jackson heart study. *MedRxiv, 7,* 1–78.

Priyadarshini, I., Mohanty, P., Kumar, R. et al. (2021). A study on the sentiments and psychology of twitter users during COVID-19 lockdown period. *Multimedia Tools and Applications.* https://doi.org/10.1007/s11042-021-11004-w

Rahman, S., Irfan, M., Raza, M., Moyeezullah Ghori, K., Yaqoob, S. & Awais, M. (2020). Performance analysis of boosting classifiers in recognizing activities of daily living. *International Journal of Environmental Research and Public Health*, *17*(3). https://doi.org/10.3390/ijerph17031082

Ramkumar, P., Thanusha, K., Soumya, U., Sahana, K. & Sushma, M. (2020). Prediction of cardiovascular disease using hybrid machine learning algorithms. *Journal of Critical Reviews*, *7*(14), 1712–1720.

Ren, Y., Fei, H., Liang, X., Ji, D. & Cheng, M. (2019). A hybrid neural network model for predicting kidney disease in hypertension patients based on electronic health records. *BMC Medical Informatics and Decision Making*, *19*(2), 51.

Sachan, S., Sharma, R. & Sehgal, A. (2021). Energy efficient scheme for better connectivity in sustainable mobile wireless sensor networks. *Sustainable Computing: Informatics and Systems*, *30*,100504.

Sachan, S., Sharma, R. & Sehgal, A. (2021). SINR based energy optimization schemes for 5G vehicular sensor networks. *Wireless Personal Communications*. https://doi.org/10.1007/s11277-021-08561-6

Sadek, R. M., Mohammed, S. A., Abunbehan, A. R. K., Ghattas, A. K. H. A., Badawi, M. R., Mortaja, M. N., Abu-Nasser, B. S. & Abu-Naser, S. S. (2019). Parkinson's disease prediction using artificial neural network. *International Journal of Academic Health and Medical Research*, *3*(1), 1–8.

Safari, S., Baratloo, A., Elfil, M. & Negida, A. (2016). Evidence based emergency medicine; part 5 receiver operating curve and area under the curve. *Emergency*, *4*(2), 111.

Schmidt-Erfurth, U., Sadeghipour, A., Gerendas, B. S., Waldstein, S. M. & Bogunović, H. (2018). Artificial intelligence in retina. *Progress in Retinal and Eye Research*, *67*, 1–29.

Sharma, R., Kumar, R., Satapathy, S. C., Al-Ansari, N., Singh, K. K., Mahapatra, R. P., Agarwal, A. K., Le, H. V. & Pham, B. T. (2020). Analysis of water pollution using different physicochemical parameters: A study of Yamuna river. *Frontiers in Environmental Science*, *8*, 581591. https://doi.org/10.3389/fenvs.2020.581591

Sharma, R., Kumar, R., Sharma, D. K., Priyadarshini, I., Pham, B. T., Bui, D. T. & Rai, S. (2019). Inferring air pollution from air quality index by different geographical areas: Case study in India. *Air Quality, Atmosphere and Health*, *12*, 1347–1357.

Sharma, R., Kumar, R., Singh, P. K., Raboaca, M. S. & Felseghi, R.-A. (2020). A systematic study on the analysis of the emission of CO, CO_2 and HC for four-wheelers and its impact on the sustainable ecosystem. *Sustainability*, *12*, 6707.

Sharma, S. et al. (2020). Global forecasting confirmed and fatal cases of COVID-19 outbreak using autoregressive integrated moving average model. *Frontiers in Public Health*. https://doi.org/10.3389/fpubh.2020.580327

Vo, M. T., Vo, A. H., Nguyen, T., Sharma, R. & Le, T. (2021). Dealing with the class imbalance problem in the detection of fake job descriptions. *Computers, Materials & Continua*, *68*(1), 521–535.

Vo, T., Sharma, R., Kumar, R., Son, L. H., Pham, B. T., Tien, B. D., Priyadarshini, I., Sarkar, M. & Le, T. (2020). Crime rate detection using social media of different crime locations and Twitter part-of-speech tagger with Brown clustering. *Journal of Intelligent & Fuzzy Systems*, *38*(4), 4287–4299.

2 AI Impacts on Photonic Crystal Sensing for the Detection of Tumors

Sunil Sharma and Lokesh Tharani

CONTENTS

INTRODUCTION

During experiments on inhibited spontaneous emission [1] in 1987 E. Yablonovitch used the term hotonic crystal. After that, several scientists used it in their experiments and designed various photonic crystal structures in one dimension (1D), two dimensions (2D) and three dimensions (3D) [1 2]. Photonic crystal fibres [3] are well-known class of optical fibres [4] and they use photonic crystals to constitute cladding [5](outer part of structure) surrounding the core [6](inner part of structure). Islam et al. [7] introduced the optical fibre sensing principle and mechanism. They introduced the notion that optical fibres are a fundamental requirement for sensing. In future these sensors will be the most widely used sensors in every domain. Due to their properties, like being immune to electromagnetic interference (EMI), their inert nature against chemical and biological changes, low transmission loss and high accuracy and fast response [7], optical fibre sensors are mostly used in various sensing applications (Figure 2.1).

Photonic crystals are low loss periodic arrangements of microscopic air holes which run along the complete length of an optical fibre. Fibre optic sensors are classified on the basis of intensity; they are spectrally and interferometric-based sensors. These sensors offer high surface area in low volume and depict extraordinary sensing

DOI: 10.1201/9781003097518-2

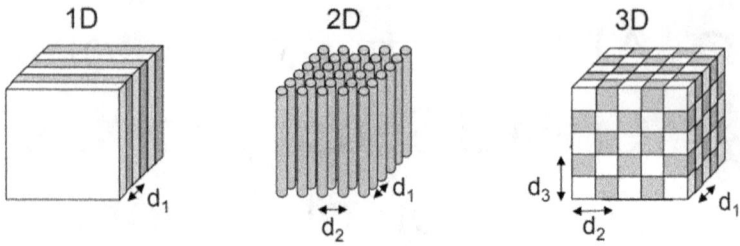

FIGURE 2.1 Photonic crystals [1].

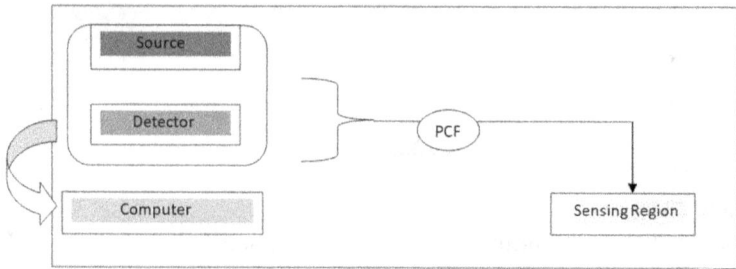

FIGURE 2.2 PCF sensing and analysing principle.

properties in the biomedical and pharma industries. A range of meta-materials [8] direct the encroachment of photonics in the design, modelling and simulating of photonic crystal fibre (PCF) structures. PCF sensors function along with a dedicated finite difference time domain (FDTD) [9] method.

The principle of sensing converts the whole setup into interferometric mode as it superimposes PCF with dedicated photonic crystal (PhC) material [10](silica glass Si, borosilicate crown glass Bk7, chalcogenide glass As_2Se_3, borofloat glass, lime soda, etc). In this way PhC performs smart sensing to obtain efficient outcomes. PCF sensors are beneficial over conventional sensors in many aspects as they offer immense design flexibility with their holey internal structure [11] (Figure 2.2).

BIOLOGICAL SENSING

Biological sensing, similar to other sensing principles, follows Bragg's law [12]. Bragg's law in combination with Snell's law is given in Figure 2.3.

$$m\lambda = 2D \ (n_{eff}^2 - \sin^2 \theta)^{1/2} \tag{2.1}$$

where
  m is the order of diffraction
  λ is the wavelength
  D is the distance between atoms in crystal
  n_{eff} is effective refractive index

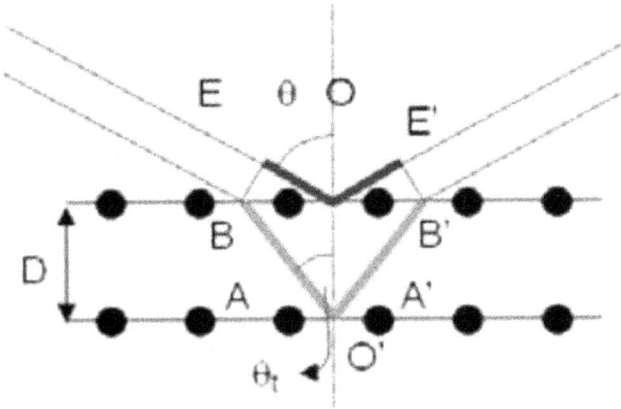

FIGURE 2.3   Bragg's and Snell's law [12].

Equation (2.1) gives that variation in distance and/or in the effective refractive index induces an alteration in the waveguide of reflected light in the sensing region [13]. It also indicates variation of the effective refractive index. The biological sensors [14] show the stimulus changes in the optical refractive index or the variation in structure. It can be seen by changing the colors and observed by reflectivity or transmittivity measurement [15]. Different kinds of hydrogels [16] can be employed for the detection of defect biomolecules. PhC sensors based on wavelength shift [17] are designed to operate in terms of sensitive to effective refractive indices n_{eff} between infected cells and normal cells [18].

This chapter is divided into four sections. The first section introduces optical fibre, photonic crystal fibre, methods, sensing principles, etc. The second section discusses tumors and their classifications. The third section covers information on the proposed dual core PCF-based sensor for the diagnosis of tumors. The fourth section is about the use of AI algorithms to detect tumors and AI impacts on detection are explained. For this purpose, various AI algorithms are used, and the optimised result is displayed.

TUMORS AND THEIR CLASSIFICATION

Tumors seem to arise when some of the genes in a cell, the chromosomes [19], are damaged, and therefore may not work appropriately. The expression of these genes is usually to regulate the rate of cell division. According to the survey report provided by the National Brain Tumor Society (NBTS) [20], more than 120 types of cancer are identified. Some tumors like glioblastoma multiforme (GBM) [21] are malevolent and may be fast-growing while the different types of tumors like meningioma, may be slow-growing and are compassionate. GBM is an aggressive type of central nervous system (CNS) [22] tumor that occurs on the surface of the brain and can be displayed in each lobe of the brain. On the other hand, meningioma is usually found in the cells which surround the brain and spinal cord [23, 24].

PROJECTED PCF-BASED REFRACTIVE INDEX SENSOR

A PCF-based dual core refractive index sensor is projected for the diagnosis of tumor cells in the human brain. Silica glass is selected as the core material for designing this structure, keeping the diameter of the air hole at 1.2 μm. Elliptical air holes are used in the first layer with the semi major and semi minor axes as 1.2 and 0.8 μm respectively. The pitch value (distance from one air hole to another) is kept at 2 μm (Figure 2.4).

To avoid the effect of reflected waves, perfect matched layer (PML) is incorporated into the projected structure. A variation in confining samples for varying refractive index modes 1.4053, 1.4055, 1.4088, 1.41... is obtained (Figure 2.5).

Samples taken from the liquid biopsy in bio fluidic form [25] is conveyed into the hole of the PCF structure for detection of its confinement through the core region. Here it is necessary to measure the refractive index of tumor samples to analyse them in an efficient way. The refractive index is used to detect variation in the tumor cells.

A refractive index-based sensor is arranged in a setup to obtain the transmission spectrum of a tumor cell and a normal cell. For this purpose, a setup is arranged and shown in Figure 2.6. It consists of an optical source (OS), optical fibres (OF), an optical spectrum analyser (OSA), a polarisation mode controller (PMC), a 3 dB coupler and an optical detector (OD) to detect the variation in the refractive indices of various cells [26]. The proposed PCF-based refractive index sensor is connected in between OF and OSA. The optical source generates a light beam which is split by the

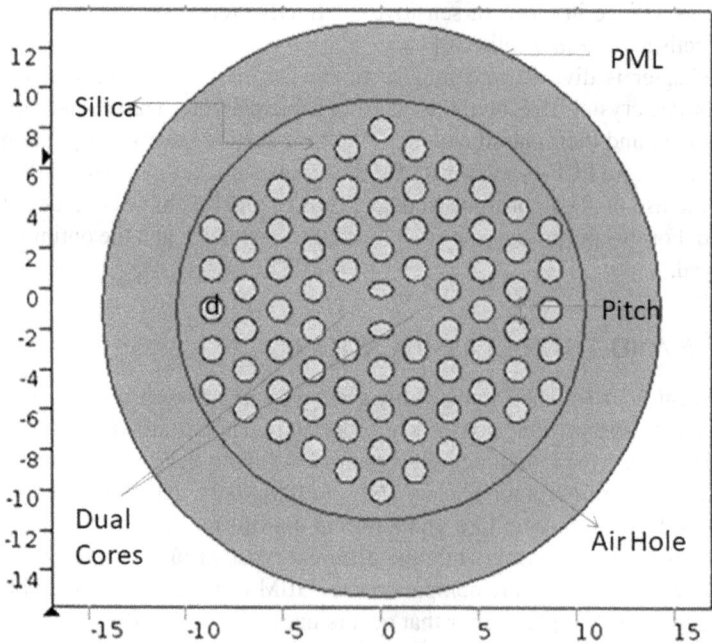

FIGURE 2.4 Proposed dual-core PCF.

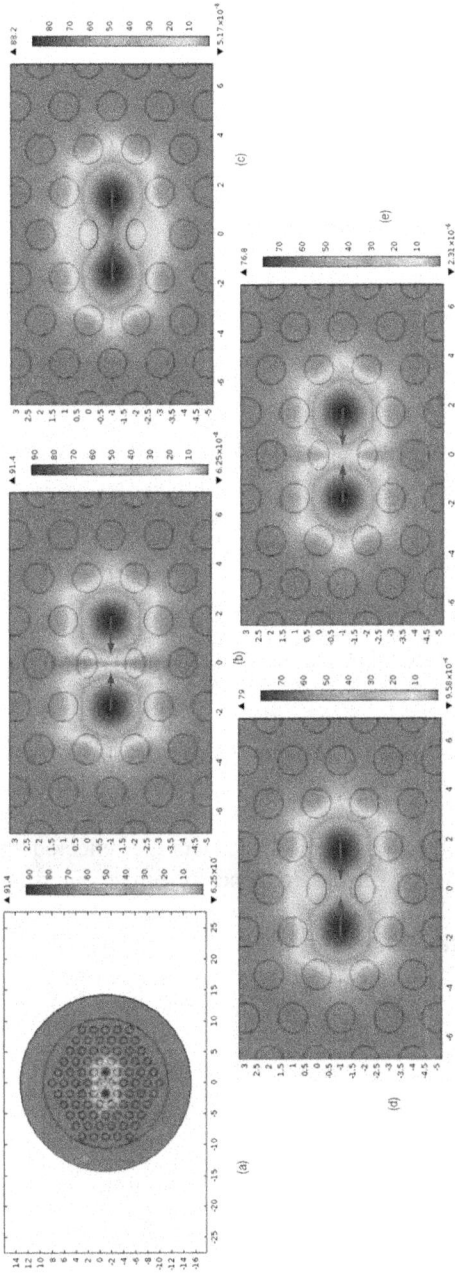

FIGURE 2.5 Samples confinement for different values of index modes. It shows variation in the refractive index and the same time variation in sample confinement is observed.

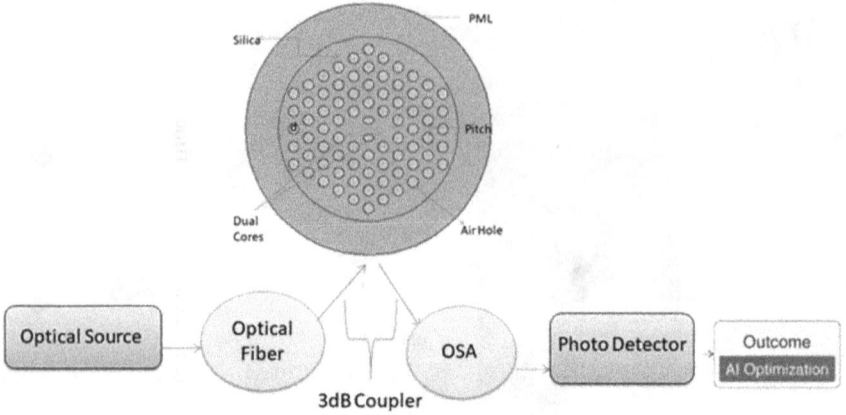

FIGURE 2.6   Setup arrangements with proposed PCF sensor.

3dB coupler. These split beams create a phase difference which can be detected and analysed. The PMC is used to adjust the polarisation [27] if it occurs during analysis.

The refractive index of silica at selected a wavelength region, i.e., from 1.5 µm to 4 µm can be calculated by Sellmier's equation [28].

i.e.,

$$n^2 - 1 = \sum \frac{Ai\lambda^2}{(\lambda - \lambda i)((\lambda + \lambda i))} \tag{2.2}$$

Where λ is the selected wavelength and n is the refractive index (Figure 2.6).

The tumor cells can be sensed by using samples of an infected cell in the air cavity [29], as the variation in the wavelength is sensed, the energy coupling and energy transfer is observed. A PML restricts the reflected stray energy [30]. The intensity of light [31] is transformed and identified as a different part of the arrangement [32].

Sensitivity can be obtained by using the equation shown below

$$r_f = f\left(\frac{n_r}{n_c}\right) \tag{2.3}$$

Where

n_c is the core refractive index (R.I.),

n_r is the RI of the fluid,

r_f is relative sensitivity coefficient and

f is the ratio of optical power within large holes to the total power [33] which is given as

$$f = \int \left[(E_x H_y - H_x E_y) \right]_{Samples} \Big/ \int \left[(E_x H_y - H_x E_y) \right]_{total} \tag{2.4}$$

ARTIFICIAL INTELLIGENCE ALGORITHM
HELPS TO DETECT TUMORS

A tumor is defined as abnormal cells [34] which cultivate in the human body. An early diagnosis is of paramount importance to improve the prognosis of patients [35], as well as thanks to new imaging techniques and the use of AI algorithms that could help doctors to identify the tumor.

Photonic integrated circuits [36] made it possible to develop the use of artificial neural networks very quickly, which is the basis for a novel class of information-based machines [37]. The algorithms implemented in hardware have the potential to meet the growing demand for AI in the detection of cancer and cancer treatment. Neuromorphic photonics [38] provide sub-nanosecond delays [39] and a possibility for further expansion of the field of AI.

Here, in this next section, from the available variety of AI techniques, some of the selected algorithms, such as support vector machine (SVM) [40], logistic regression (LR), fuzzy logic (FL) [41], random forest (RF), artificial neural network (ANN) and k-nearest neighbor (kNN) [42] are presented in this chapter.

Support vector machine (SVM)is an asymmetrical identification of structure-based minimisation [43] of possibility and a classification of linear and non-linear statistics datasets. The intention is to discern an agitated hydroplane within an N-dimensional space (where N is the number of attributes) that predominantly classifies the statistics data points. By means of these support vectors it exploits the edges of the classifier [44]. It shows accuracy of about 96%.

Logistic regression (LR) is a commanding and well-entrenched technique [45] for supervised classification. Logistic regression makes use of a five-nearest neighbors or a ten-nearest neighbors. Aggregation coalesces the design of the assembly learning[46] and bootstrap aggregation, or bagging, and, as previously mentioned, allows for additional results. Bootstrap samples from a training dataset, and the end-of-the-box samples are used to train a select group of students. To make use of LR in terms of a binary classifier, a porch has to be allocated to differentiate two classes. It shows accuracy of about 88%.

Fuzzy logic (FL) sets employ diverse rules for mean standard deviation (MSD) [47] of attribute values and it provides accuracy of around 93%.

The random forest (RF) is a group classifier [48] and consists of numerous decision trees related to the sense that a forest is a collection of countless trees [49]. An RF algorithm believes the outcomes from numerous special decision trees can decrease the discrepancy resulting from the deliberation of a single decision tree for a similar dataset [50, 54]. The accuracy obtained is around 90%.

An ANN presents a system for the diagnosis [51] and classification of tumors from MRIs [52] by means of a back-propagation network and produces accuracy rates like 77.56%, 72.5% and 95%.

k-nearest neighbors (kNN) is a simple, straightforward classifier which can provide good performance for the optimal values of k. In addition, kNN training is very fast [53], and every game mission is very simple. kNN is used in the magnetic resonance images and showed the lowest detection level of 70%, compared to the number of neighbors chosen as 7 (Figures 2.7 and 2.8).

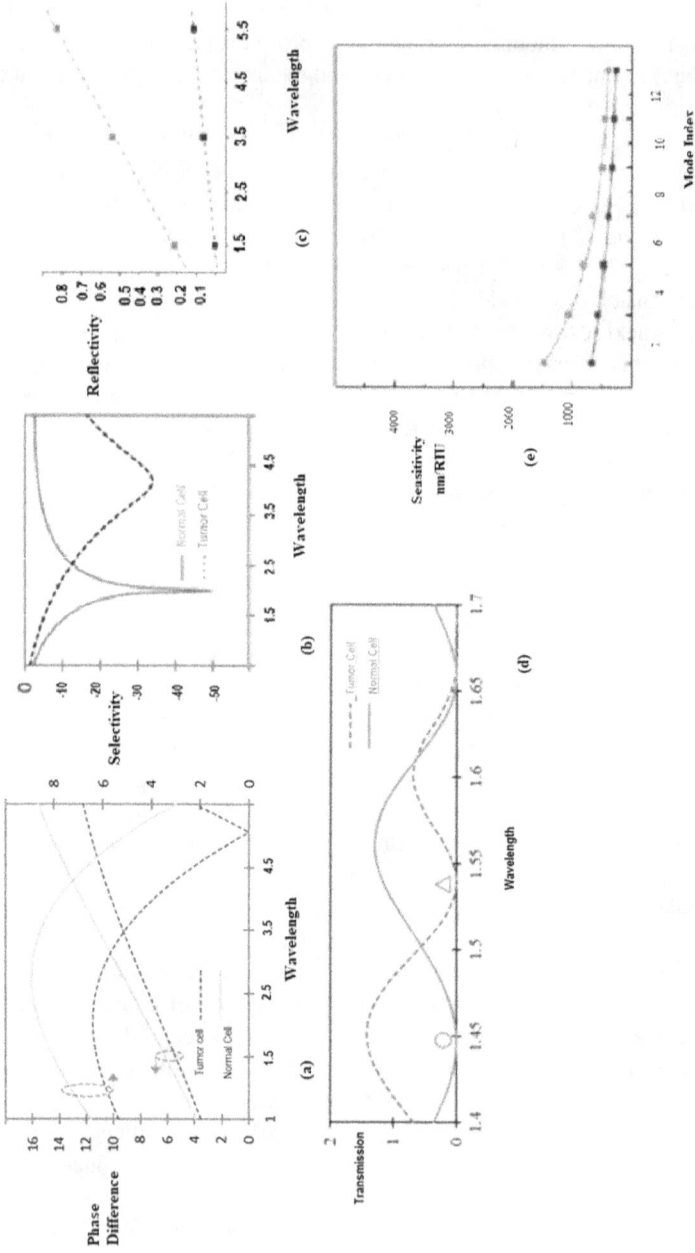

FIGURE 2.7 Parameters (a) phase difference, (b) sensitivity, (c) reflectivity, (d) transmission and (e) sensitivity measured with wavelength variation for arranged setup and proposed PCF.

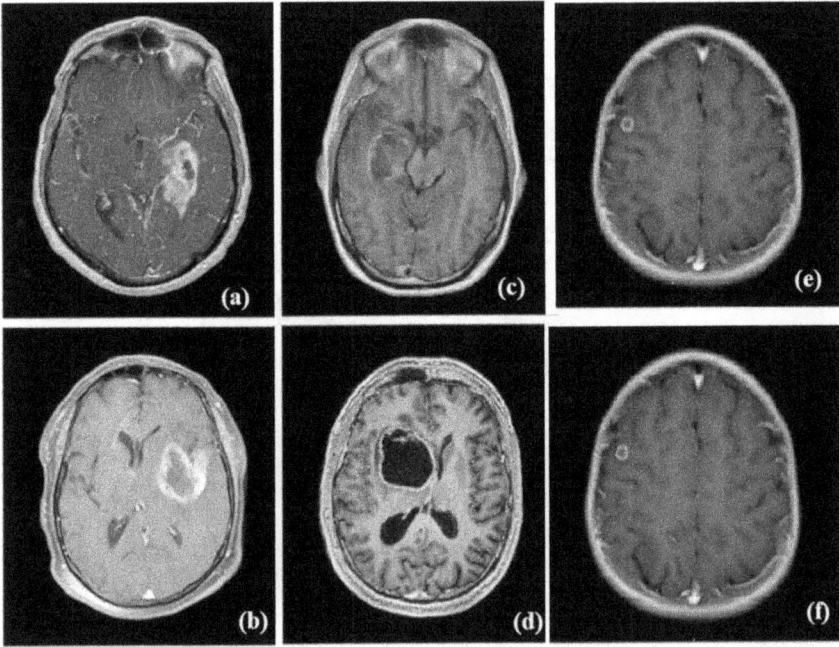

FIGURE 2.8   Impact observed using AI algorithms on tumor detection.

After applying these above-mentioned algorithms to all images in the database, we compute the sensitivity as:

$$\text{Sensitivity} = \frac{\text{Identified true positive}}{\text{Real no. of positiveses}}$$

RESULT AND DISCUSSION

The aim of designing a dual-core PCF-based refractive index sensor is to achieve high sensitivity towards tumor detection. For this purpose, the proposed design is arranged in a manner in which a tumor cell cavity is infiltrated in the cladding region. To achieve optimised sensitivity various AI algorithms have been classified on the basis of suspicious region detection. It has been observed that various optical characteristics like refractive index, phase difference, transmissivity and loss can be used to sense infiltration values between normal cells and tumor cells. AI algorithms are used to find out the accuracy, sensitivity and selectivity on the basis of RI changes for normal cells and tumor cells. The maximum sensitivity observed was 32,358 nm/RIU and the minimum sensitivity observed as 11,258 nm/RIU. On the other hand, the highest accuracy reported for SVM was 96% and minimum accuracy with kNN was 70%. The RI for tumor cells varies from 1.3342 to 1.4251. The number of layers can be varied for the variation in the results obtained. The transmission spectrum

TABLE 2.1

Refractive Index, Sensitivity, Accuracy and Concentration of Normal Cells and Tumor Cells

	Concentration (%)		RI		Sensitivity	Accuracy
AI Algorithm	GBM	Meningioma	Normal Cell	Tumor Cell	(nm/RIU)	(%)
SVM	76.30	57.20	1.38	1.4251	32,358	96%
LR	67.83	42.74	1.33	1.3955	30,534	88%
FL	71.48	49.36	1.34	1.3924	31,435	93%
RF	69.35	48.71	1.36	1.3521	31,078	90%
KNN	50.48	41.58	1.32	1.3437	11,258	70%
ANN	73.37	52.36	1.35	1.3342	32,045	95%

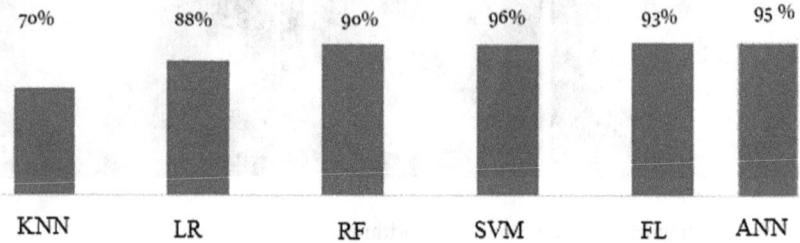

FIGURE 2.9 Performance of various AI techniques adopted for tumor detection.

represents that the wavelength of significant peak is enhanced with increasing the RI of tumor cells.

Table 2.1 indicates the various RI values, accuracy and sensitivity of tumor cells and normal cells along with identified AI algorithms.

Based on these investigations performed by mean of identified AI techniques between different classes of tumors, the computer illustrates the apprehensive region or laceration and calculates approximately its probability of occurrence in the identified region which is presented in Figure 2.9.

It defines that GBM tumors have more concentration compared to meningioma. The various concentration values observed by different AI algorithms are presented with sensitivity and accuracy variation in Figure 2.10.

CONCLUSION

Photonic crystals (PhC) have been utilised due to their generous structures and flexibility to adapt to every field. A dual-core PCF RI sensor is designed with selected diameter, pitch value and silica glass to achieve the identified goal. The goal is to improve performance, like sensing ability, accuracy and selectivity by using identified AI algorithms. kNN, SVM, ANN, FL, RF and LR are used to observe sensing

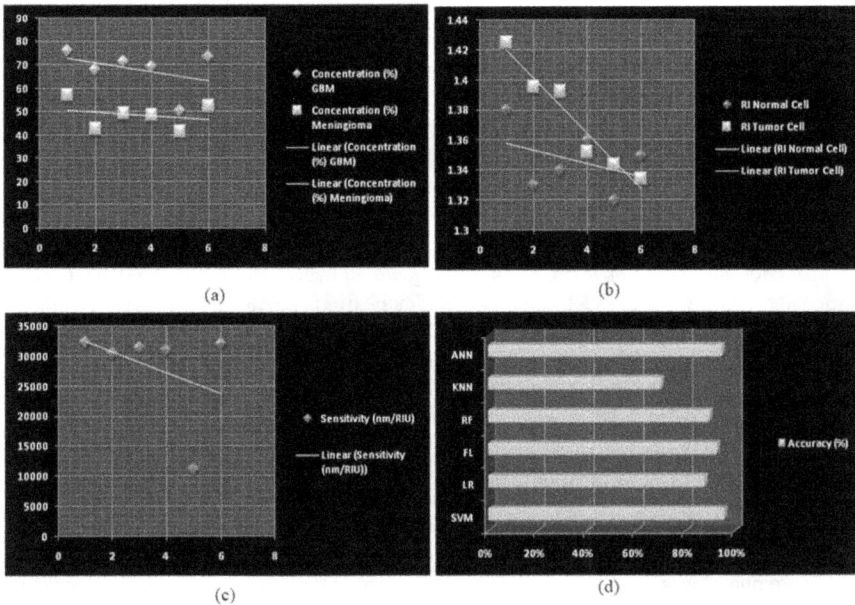

(a)　　　(b)

(c)　　　(d)

FIGURE 2.10 Analysed outcomes (a) concentration, (b) refractive index, (c) sensitivity and (d) accuracy of normal and tumor cells by means of various AI algorithms.

ability and by selecting these algorithms' accuracy and concentration of normal cells, GBM and meningioma are detected. It is concluded and mentioned above that tumor cells have refractive indices between 1.3342 and 1.4251. The maximum sensitivity observed as 32358, nm/RIU and the minimum sensitivity observed as 11258, nm/RIU. The highest accuracy reported for SVM was 96% and minimum accuracy with kNN was 70%. In summary, the PhC, where it is working as a sensor for the detection of tumors, achieves its outcomes on the basis of RI. But with the integration of AI, they perform in a much better way and produce better outcomes in terms of accuracy, sensitivity, concentration, etc. So, it is concluded that AI is better at sensing and detecting tumor cells.

In future the selection of PhC material, structure size and parameter selection can be varied along with PCF sensors to get different outcomes in a different manner.

ACKNOWLEDGEMENT

We would like to acknowledge direct and indirect resources those who have supported us for compiling datasets and obtaining outcomes.

DECLARATION

No funding is associated with this article.

CONFLICT OF INTEREST

The authors declare no conflict of interest.

REFERENCES

1. Yablonovitch, E. (1987). Inhibited spontaneous emission in solid-state physics and electronics. *Physical Review Letters*, *58*, 2059.
2. Fakhruldeen, H. F., Zahid, A. Z., Jaafar, R. M. & Abdulkareem, A. (2019). An overview of photonic crystal fiber (PCF). *Indian Journal of Natural Sciences*, *9*(53), 0976–0997.
3. Sahay, G., Sharma, S., Mathur, A. & Dhabhai, H. (2018). Designing of rectangular photonic crystal fiber to minimize the total dispersion. *International Journal of Engineering and Technology*, *7*(5), 268–270.
4. Arif, M. F. H., Ahmed, K., Asaduzzaman, S. & Azad, M. A. K. (2016). Design and optimization of photonic crystal fiber for liquid sensing applications. *Photonic Sensors*, *6*(3), 279–288.
5. Hao, R., Li, Z., Sun, G., Niu, L. & Sun, Y. (2013). Analysis on photonic crystal fibers with circular air holes in elliptical configuration. *Optical Fiber Technology*, *19*(5), 363–368.
6. Islam, M. R., Hossain, M. A., Talha, K. M. A. & Munia, R. K. (2020). A novel hollow core photonic sensor for liquid analyte detection in the terahertz spectrum: Design and analysis. *Optical and Quantum Electronics*, *52*(9), 415. https://doi.org/10.1007/s11082 -020-02532-0
7. Sharma, S. & Tharani, L. (2021). *AI for photonics and photonics for AI, material and characteristics integration*. IntechOpen, Fiber Optics. ISBN 978-1-83969-627-5.
8. Sharma, S., Tharani, L. & Sharma, R. K. (2020). Designing a nonlinear tri core photonic crystal fiber for minimizing dispersion and analyzing it in various sensing applications. In G. Mathur, H. Sharma, M. Bundele, N. Dey & M. Paprzycki (Eds), *International Conference on Artificial Intelligence: Advances and Applications 2019. Algorithms for Intelligent Systems.* https://doi.org/10.1007/978-981-15-1059-5_3
9. Tejiyot, H. K., Sharma, S., Mathur, A. & Dhabhai, H. (2018). Design to analyze low loss zero dispersion in water filled silica photonic crystal fiber. In *2018 2nd International Conference on Micro-Electronics and Telecommunication Engineering (ICMETE)* (pp. 197–202). https://doi.org/10.1109/ICMETE.2018.00051
10. Sharma, S. & Vyas, K. (2013). Novel design of honeycomb photonic crystal fiber with nearly zero flattened chromatic dispersion. In *IEEE International Conference on Communication and Signal Processing- (ICCSP'13)*. ISSN: 2315–4535.
11. Comoretto, D. (2010). *Introduction to photonic crystals: Nanostructured materials to manipulate light propagation & harvesting*. Trieste 8 – 19 February 2010.
12. Troia, B., Paolicelli, A., De Leonardis, F. & Passaro, V. M. N. (2012). *Photonic crystals for optical sensing: A review*. IntechOpen. https://doi.org/10.5772/53897
13. Sharma, S. & Tharani, L. (2021, June 10). Photonics for AI and AI for photonics: Material and characteristics integration [online first]. IntechOpen. https://doi.org/10 .5772/intechopen.97781. https://www.intechopen.com/online-first/photonics-for-ai -and-ai-for-photonics-material-and-characteristics-integration
14. Fini, J. M. (2004). Microstructure fibres for optical sensing in gases and liquids. *Measurement Science and Technology*, *15*(6), 1120–1128.
15. Park, J., Lee, S., Kim, S. & Oh, K. (2011). Enhancement of chemical sensing capability in a photonic crystal fiber with a hollow high index ring defect at the center. *Optics Express*, *19*(3), 1921–1929.

16. Olyaee, S., Naraghi, A. & Ahmadi, V. (2014). High sensitivity evanescent-field gas sensor based on modified photonic crystal fiber for gas condensate and air pollution monitoring. *Optik*, *125*(1), 596–600.
17. Patel, A., Fisher, J, Nichols, E. et al. (2019). Global, regional, and national burden of brain and other CNS cancer, 1990–2016: A systematic analysis for the Global Burden of Disease Study 2016. *The Lancet Neurology*. https://www.thelancet.com/journals/laneur/article/PIIS1474-4422(18)30468-X/fulltext#%20
18. Hollon, T., Pandian, B. & Orringer, D. (2019). Near real-time intraoperative brain tumor diagnosis using stimulated Raman histology and deep neural networks. *Nature Medicine*. https://www.nature.com/articles/s41591-019-0715-9
19. Zheng, L. & Chan, A. K. (2001). An artificial intelligent algorithm for tumor detection in screening mammogram. *IEEE Transactions on Medical Imaging*, *20*(7), 0278–0062/01.
20. Rajesh Chandra, G. & Rao, K. R. H. (2016). *Tumor detection in brain using genetic algorithm*. ICCCV 2016. https://doi.org/10.1016/j.procs.2016.03.058
21. Aslam Mollah, Md., Usha, R. J., Tasnim, S. & Ahmed, K. (2020). Detection of cancer affected cell using Sagnac interferometer based photonic crystal fber refractive index sensor. *Optical and Quantum Electronics*, *52*, 421. https://doi.org/10.1007/s11082-020 -02542-y
22. Martin, L. J. (2019). *Medically reviewed meningioma*. December 2019.
23. Dlamini, Z., Francies, F. Z., Hull, R. & Marima, R. (2020). Artificial intelligence (AI) and big data in cancer and precision oncology. *Computational and Structural Biotechnology Journal*, *18*, 2300–2311. https://doi.org/10.1016/j.csbj.2020.08.019
24. Nouman, W. M., Abd El Ghany, S. E.-S., Sallam, S. M., Dawood, A.-F. B. & Aly, A. H. (2020). Springer Science+Business Media, LLC, part of Springer Nature, 2020.
25. Sa, D. U., Fua, C. Y., Soha, K. S., Bhuvaneswari, R., Kumar, A. & Olivoa, M. (2012). Highly sensitive SERS detection of cancer proteins in low sample volume using hollow core photonic crystal fiber. *Biosensors and Bioelectronics*, *33* (1) 293–298.
26. Aslam Mollah, Md., Usha, R. J., Tasnim, S. & Ahmed, K. (2020). Detection of cancer affected cell. Springer Science+Business Media, LLC, part of Springer Nature
27. Ames, E., Canter, R. J., Grossenbacher, S. K., Mac, S., Chen, M., Smith, R. C. & Murphy, W. J. (2015). NK cells preferentially target tumor cells with a cancer stem cell phenotype. *The Journal of Immunology*, *195*(8), 4010–4019. https://doi.org/10.4049/jimmunol.1500447
28. Jain, A., Sharma, R. K., Agarwal, V. & Sharma, S. (2020). A new design of equiangular circular cum elliptical honeycomb photonic crystal fiber. In G. Ranganathan, J. Chen & Á. Rocha (Eds), *Inventive communication and computational technologies. Lecture notes in networks and systems*, vol 89. Springer. https://doi.org/10.1007/978-981-15 -0146-3_6
29. Ademgil, H. (2014). Highly sensitive octagonal photonic crystal fiber based sensor. *Optik: International Journal for Light and Electron Optics*, *125*(20), 6274–6278.
30. Sharma, S., Sharma, R. K., Gupta, R. & Dash, P. (2020). Design and analysis of elliptical core spiral silica photonic crystal fiber with improved optical characteristics. In A. Kumar & S. Mozar (Eds), *ICCCE 2019. Lecture notes in electrical engineering*, vol 570. Springer. https://doi.org/10.1007/978-981-13-8715-9_9
31. Kaur, V. & Singh, S. (2018). Performance analysis of multichannel surface plasmon resonance sensor with dual coating of conducting metal oxide. *Journal of Nanophotonics*, *12*(1), 016012.
32. Gopal, N. N. & Karnan, M. (2010, December). Diagnose brain tumor through MRI using image processing clustering algorithms such as Fuzzy C Means along

with intelligent optimization techniques. In *IEEE International Conference on Computational Intelligence and Computing Research (ICCIC)* (pp. 1–4).

33. Najadat, H., Jaffal, Y., Darwish, O. & Yasser, N. (2011). A classifier to detect abnormality in CT brain images. In *The 2011 IAENG International Conference on Data Mining and Applications* (pp. 374–377), March 2011.

34. Othman, M. F. & Basri, M. A. M. (2011, January). Probabilistic neural network for brain tumor classification. In *Second International Conference on Intelligent Systems, Modelling and Simulation (ISMS)* (pp. 136–138), January 2011.

35. Specht, D. F. (1990). Probabilistic neural networks. *Neural Networks*, *3*(1), 109–118.

36. Piccinotti, D., MacDonald, K. F., Gregory, S. A., Youngs, I. & Zheludev, N. I. (2020). Artificial intelligence for photonics and photonic materials, IOP Publishing Ltd. *Reports on Progress in Physics*, *84*, 012401.

37. Wang, J. J., Xu, Y. Z., Mazzarello, R., Wuttig, M. & Zhang, W. (2017). A review on disorder-driven metal-insulator transition in crystalline vacancy-rich GeSbTe phase-change materials. *Materials (Basel, Switzerland)*, *10*(8), 862. https://doi.org/10.3390/ma10080862

38. Soler, M., Estevez, M. C., Rubio, M. C., Astua, A. & Lechuga, L. M. (2020). How nano photonic label-free biosensors can contribute to rapid and massive diagnostics of respiratory virus infections: COVID-19 case. *ACS Sensors*, *5*(9), 2663–2678. https://doi.org/10.1021/acssensors.0c01180

39. Begam, S., Vimala, J,., Selvachandran, G., Ngan, T. T. & Sharma, R. (2020). Similarity measure of lattice ordered multi-fuzzy soft sets based on set theoretic approach and its application in decision making. *Mathematics*, *8*, 1255.

40. Vo, T., Sharma, R., Kumar, R., Son, L. H., Pham, B. T., Tien, B. D., Priyadarshini, I., Sarkar, M. & Le, T. (2020). Crime rate detection using social media of different crime locations and Twitter part-of-speech tagger with Brown clustering. *Journal of Intelligent & Fuzzy Systems*, *38*(4), 4287–4299.

41. Nguyen, P. T., Ha, D. H., Avand, M., Jaafari, A., Nguyen, H. D., Al-Ansari, N., Van Phong, T., Sharma, R., Kumar, R., Le, H. V., Ho, L. S., Prakash, I. & Pham, B. T. (2020). Soft computing ensemble models based on logistic regression for groundwater potential mapping. *Applied Sciences*,*10*, 2469.

42. Jha, S. et al. (2019). Deep learning approach for software maintainability metrics prediction. *IEEE Access*, *7*, 61840–61855.

43. Sharma, R., Kumar, R., Sharma, D. K., Priyadarshini, I., Pham, B. T., Bui, D. T. & Rai, S. (2019). Inferring air pollution from air quality index by different geographical areas: Case study in India. *Air Quality, Atmosphere and Health*, *12*, 1347–1357.

44. Sharma, R., Kumar, R., Singh, P. K., Raboaca, M. S. & Felseghi, R.-A. (2020). A systematic study on the analysis of the emission of CO, CO_2 and HC for four-wheelers and its impact on the sustainable ecosystem. *Sustainability*, *12*, 6707.

45. Sharma, S. et al. (2020). Global forecasting confirmed and fatal cases of COVID-19 outbreak using autoregressive integrated moving average model. *Frontiers in Public Health*. https://doi.org/10.3389/fpubh.2020.580327

46. Malik, P. et al. (2021). Industrial internet of things and its applications in industry 4.0: State-of the art. *Computer Communication*, Elsevier, *166*, 125–139.

47. Sharma, R., Kumar, R., Satapathy, S. C., Al-Ansari, N., Singh, K. K., Mahapatra, R. P., Agarwal, A. K., Le, H. V. & Pham, B. T. (2020). Analysis of water pollution using different physicochemical parameters: A study of Yamuna river. *Frontiers in Environmental Science*, *8*, 581591. https://doi.org/10.3389/fenvs.2020.581591

48. Dansana, D., Kumar, R., Parida, A., Sharma, R., Adhikari, J. D. et al. (2021). Using susceptible-exposed-infectious-recovered model to forecast coronavirus outbreak. *Computers, Materials & Continua*, *67*(2), 1595–1612.

49. Vo, M. T., Vo, A. H., Nguyen, T., Sharma, R. & Le, T. (2021). Dealing with the class imbalance problem in the detection of fake job descriptions. *Computers, Materials & Continua, 68*(1), 521–535.

50. Sachan, S., Sharma, R. & Sehgal, A. (2021). Energy efficient scheme for better connectivity in sustainable mobile wireless sensor networks. *Sustainable Computing: Informatics and Systems, 30*, 100504.

51. Ghanem, S., Kanungo, P., Panda, G. et al. (2021). Lane detection under artificial colored light in tunnels and on highways: an IoT-based framework for smart city infrastructure. *Complex & Intelligent Systems*. https://doi.org/10.1007/s40747-021-00381-2

52. Sachan, S., Sharma, R. & Sehgal, A. (2021). SINR based energy optimization schemes for 5G vehicular sensor networks. *Wireless Personal Communications*. https://doi.org/10.1007/s11277-021-08561-6

53. Priyadarshini, I., Mohanty, P., Kumar, R. et al. (2021). A study on the sentiments and psychology of twitter users during COVID-19 lockdown period. *Multimedia Tools and Applications*. https://doi.org/10.1007/s11042-021-11004-w

54. Azad, C., Bhushan, B., Sharma, R. et al. (2021). Prediction model using SMOTE, genetic algorithm and decision tree (PMSGD) for classification of diabetes mellitus. *Multimedia Systems*. https://doi.org/10.1007/s00530-021-00817-2

3 Cybersecurity Solutions and Communication Technologies for Internet of Things Applications

Gagan Varshney and Bharat Bhushan

CONTENTS

DOI: 10.1201/9781003097518-3

INTRODUCTION

The Internet of Things (IoT) has developed a new emerging model in which devices are able to communicate and collaborate with each other. The adaption of the IoT is driving new innovations in industry [1]. Application areas of the IoT include smart home, smart healthcare, intelligent transportation and smart grids. These are vulnerable to a wide range of security issues like finance, compliance and business operations. IoT security may be challenging due to the diverse, complex and dynamic nature of devices [2]. An ever-increasing number of cyberattacks on IoT systems have influenced both people and organisations. According to a survey, about 81% of organisations have experienced cyberattacks in recent years [3]. However, the survey also finds that about 26% of the organisations did not use any security technologies. Cybersecurity needs to be implemented on the IoT to reduce the risk to privacy of people as well as organisations. Cybersecurity is the field that suggests compliance and implementation needs to be applied to IoT devices so that secure communication can be effected over the IoT network. Commercial IoT devices are used in applications such as location monitoring, live video recording and building access control [4, 5]. In order to make such devices more reliable, security should be considered at the early stage in the design of new products [6]. But this security challenge is further complicated due to limits in terms of power, storage and computation capabilities. Such limitations prevent the chances of adapting standard security mechanisms in IoT devices [7]. These devices operate in isolated systems that are less vulnerable to security threats. Hence, the manufacturer used low-cost sensors and actuators. Initially when these devices were designed, the manufacturer did not possess cybersecurity knowledge and might not have been aware of the risks associated with devices when they are connected to the internet or global network. There is a need to provide up-to-date information on the current status of IoT cybersecurity.

The major security challenges with the IoT are inadequate testing and updating. These issues create vulnerabilities in the IoT. Hence, IoT devices must be tested properly before launching globally and need to be updated regularly, but due to the complex structure of the IoT, users expect that security analysis can be done automatically at regular intervals without human intervention. This expectation can be fulfilled by using intelligence learning techniques. These techniques have to be deployed on the IoT network to analyse behavior and identify anomalies in an efficient way. This technique is formed by considering four main principles of learning theory: probabilistic methods, fuzzy logic, neural networks and evolutionary computing. Intelligence-enabled techniques enhance cyber defence capabilities and can help to make an efficient intrusion detection system (IDS).

In the literature, there are limited studies about the solution to security issues on the IoT network. To fill this gap, this chapter elaborates possible security solutions for the emerging IoT and how various existing mechanisms can be used in an efficient way to improve the security of the IoT. The major contributions of this work are enumerated as follows.

- This chapter presents a detailed discussion relating to the background of cybersecurity and security challenges in a typical IoT ecosystem.

- This chapter describes the cybersecurity issues and solutions associated with each layer of the IoT framework.
- This chapter describes different standard security mechanisms for IoT services and highlights the security scenarios of various IoT communication protocols and standards.
- Finally, this chapter highlights new emerging automated security techniques based on computational intelligence for the IoT network.

The remainder of the chapter is organised as follows. The next section presents the role of cybersecurity in IoT architecture and highlights the security issues (and their proposed solutions) at each layer of IoT architecture. The following section covers the main security mechanisms for IoT services and is followed by a section discussing the security concerns of major IoT communication protocols. The final section describes the role of computational intelligence towards realisations of a secured IoT ecosystem and is followed by a conclusion.

CYBERSECURITY IN IOT ARCHITECTURE

The IoT is a network that is connected through various networking devices, and these networking devices are wireless; therefore to maintain the complexity of network, a layered architecture model is required. Every level of the architecture has its own security concerns, and these layers are linked with each other in order to process the information, so a security strategy needs to be designed for the entire system [8]. In 2018, there were about 23 billion devices connected, which was more than double the world population. According to the survey there will be over 80 billion devices by 2025. Hence, there is need for cybersecurity to ensure the security of interconnected IoT devices and services from unauthorised access. Security should be designed in such a way so that information or data can be exchanged securely within devices and externally. A literature survey of cybersecurity suggested three common problems with the IoT network that should be focused on are data confidentiality, trust and privacy. Cybersecurity in the IoT needs to implemented to enhance device security, data security and individuals' privacy. IoT architecture consists of five layers and cybersecurity focuses on security issues and solutions for each layer.

CYBERSECURITY AT THE PERCEPTION LAYER

The perception layer is the primary layer of the IoT framework that collects meaningful data from its surroundings, like moisture content, temperature, noise, etc. by using a number of sensing devices. This information passes to another layer so that some action can be taken based on it. IoT devices are recurrently receiving massive amounts of information from their surroundings and therefore several techniques are needed to save energy or to reduce energy consumption. Technologies like machine learning are generally used to make a reliable interpretation based on the data received [9]. But the devices used in the IoT have limited ability and these devices are often called resource-constrained devices, due to which the inner computation-intensive security built into these IoT devices has been vulnerable [10, 11]. At the perception level, the more vulnerable issue is the cloning of device chips. For example, replicas of radio frequency identification (RFID) labels can be used to implement attacks like distributed denial-of-service (DDoS) attacks. Therefore, to enhance security, physical unclonable functions (PUFs) ease the overall authentication and identification. But due to the limited ability of IoT devices we cannot use PUF chips directly, therefore researchers have designed lightweight PUF chips for such resource-constrained devices. As PUF chips are not cloneable, there might be a chance to clone a PUF key once it is extracted. Therefore, several valid protocols have been suggested to secure PUF chips [12, 13].

Cybersecurity at the Network Layer

This plays a crucial role in maintaining the entire performance of IoT security, since information should be securely exchanged for the functioning of connected nodes and the IoT framework. A security technique called an intrusion detection system (IDS) detects anomalies, monitors packets and takes security measures [14]. The IDS installs numerous intrusion detection methods: protocol verification for categorising suspicious acts, evolutionary algorithms to identify intrusions based on certain faults, attempted intrusions and behavior [15], statistical analysis [16], deep learning for categorising network breach patterns [17] and data mining methods like the random forest method [18]. One of the optimistic methods is deep learning models which show optimistic outcomes in detection of DDoS attacks with an accuracy of about 97% [19]. Hybrid techniques use classification methods and feature reduction for detecting anomalies over the IoT networks because of the increasing adaption of IoT applications for future growth.

Cybersecurity at the Processing Layer

The processing layer is responsible for storage, processing, computation, etc. and this level stores all the information received and shares suitable information with devices based on the device's identity and address. Emerging technologies like fog computing and cloud computing are widely used for storing and processing the significant amount of data bundles received from connected devices in the IoT network simultaneously. In fog computing, IDS detects anomalies at fog terminals. But IDS alone is not enough to maintain security, therefore an advance hybrid method designed by using a combination of IDS, Markov models and virtual honeypot devices (VHD) is used to enhance threat detection, process and identify malicious devices. This hybrid method also reduces chances of false alarms. Modern technology like blockchain can be used at the processing layer to create IoT security certificates for devices to be installed safely and in an automated way. Various emerging application areas of blockchain in IoT include healthcare, smart energy and supply chains [20, 21]. Together blockchain and the IoT can allow both secure access and exchange IoT data within their private network without the involvement of any centralised control and management system.

CYBERSECURITY AT THE APPLICATION LAYER

This layer interacts with all application processes based on the data received from the processing layer. This layer performs functions like turning on or off a device, alert alarm, broadcast information, sending mails, etc., various activities like business analytics and big data, monitoring and control, and sharing and collaboration are trending IoT applications [22]. Every application area such as smart home or smart vehicle has different security management methods [23, 24]. Some security issues at this layer involve the security of protocols such as XMPP (Extensible Messaging and Presence Protocol), MQTT (MQ Telemetry Transport) and CoAP (Constrained Application Protocol) [25], insufficient audit tools [26], inadequate authentication [27] and improper patch management [28]. Various solutions have been suggested for this problem such as private information protection, heterogeneous network authentication, key management and access control [29].

CYBERSECURITY AT THE SERVICE MANAGEMENT LAYER

This deals with the human and organisational view of cybersecurity. Trust and confidentiality are the security issues that can influence the use of IoT services and applications. Threats to privacy and security become common in the use of cloud-based services. It is necessary to protect the privacy of those devices which are involved in the processing of personal information [30]. The integration of privacy safeguards at an early stage of the IoT not only build confidence but promote the adoption of IoT services globally. However, it is not easy to protect the privacy of IoT devices as they are resource-constrained devices or use lightweight technologies for their operations [31]. For privacy protection in smart healthcare, a privacy protector framework is designed to defend against cyberattacks such as data leakage and collusion attacks.

The five layers of IoT architecture are depicted in Figure 3.1.

SECURITY MECHANISMS FOR IOT SERVICES

In this section, we present different security mechanisms that can be used to reduce the risk of cyberattacks over the IoT network. The main security mechanisms for IoT services are explored in the subsections below.

ENCRYPTION

Encryption is the security technique to ensure confidentiality and integrity during transmission. It is a process of changing the original content (plaintext) into an unintelligible one (ciphertext) using a number of encryption methods or a hash function that can be easily decrypted (decoded) by a secret key only. After the encryption process, an attacker can only access encrypted text (ciphertext), but will not be able to decode the original message. The encryption can be of two types, symmetric key encryption and asymmetric key encryption, and this classification is done on the basis of the type of keys used for the encryption and decryption process. In

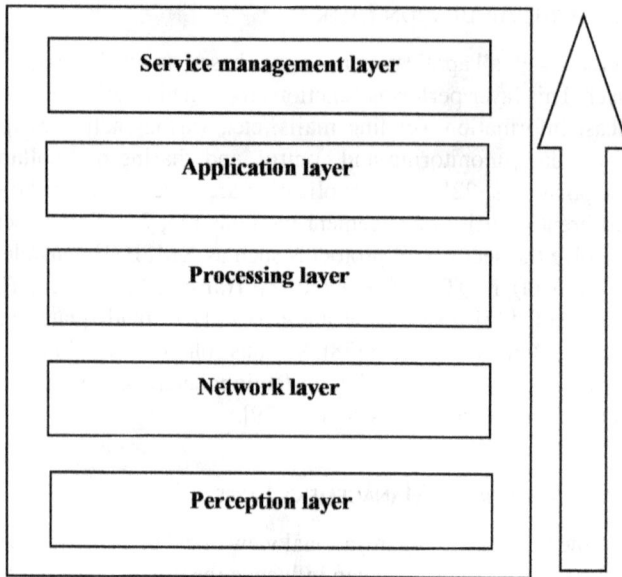

FIGURE 3.1   General IoT architecture.

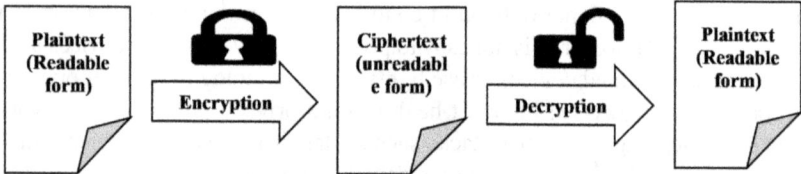

FIGURE 3.2   General mechanism of encryption and decryption.

symmetric key encryption, a common key is used for the encryption and decryption, but this technique is not assumed to be secure because the sender has to send the key along with the message, therefore it requires a secure path for sending the key. The asymmetric key encryption technique is more secure than the symmetric one because it uses the concept of public and private keys. A public key is used for the encryption and this key may be available globally, but a private key should be hidden because it is used for the decryption process [32]. Hence, to assure confidentiality during communication, the message is encrypted at the sender's end by using the receiver's public key which can then be decrypted by using their own private key as depicted in Figure 3.2.

STANDARD ENCRYPTION MECHANISM

An encryption mechanism can be divided into two different types: stream cipher, in which the plaintext is encrypted bit-by-bit (or byte-by-byte) and block cipher, in which the block of plain text is encrypted to produce ciphertext of the same length. Block cipher can be operated in various modes: cipher block chaining (CBC), electronic codebook (ECB), output feedback (OFB) and cipher feedback (CFB). Advanced encryption standard (AES) is a symmetric encryption mechanism that falls under the category of block cipher. In AES, a combination of substitution, transposition and a linear cipher is used in a repetitive manner to achieve final encryption, depending upon the number of bits of key AES, classified as AES-128, AES-192, AES-256. There are also asymmetric key cryptosystems like Rivest Shamin Adleman (RSA) [33], the McEliece [34] and the ElGamal [35] algorithms. Not only the confidentiality but also the integrity and authentication can be provided by the encryption process: digital signatures (asymmetric mechanisms), message authentication PIN (symmetric mechanisms) and hash functions. But in the IoT additional mechanisms are required with these functionalities because IoT services require the broadcasting of messages to many nodes. Therefore, the concept of tags is introduced, a tag is attached with plaintext that identifies the sender and verifies its authenticity using a trusted receiver, so without wasting time and resources all nodes can just read that message [36, 37].

LIGHTWEIGHT CRYPTOGRAPHY

Increasing the growth of the IoT containing many constrained devices, a research committee has tried to develop a modified security solution called lightweight cryptography for such devices constrained by energy and resources. Lightweight cryptography is new field of cryptography that is designed to operate in a constrained environment. In lightweight cryptography, newly modified block cipher, stream cipher, message authentication and hash function have been suggested so that cryptographic techniques can be executed by such low-complexity devices with limited resources. Verified standards contained modified encryption techniques [38], the block ciphers CLEIFA [39] and PRESENT [40]. PHOTON [41] and SPONGENT [42] are the two modified hash functions proposed by the organisations in 2012. In 2013, many projects and research were started by NIST (National Institute of Standards and Technology) to develop a lightweight cryptography solution for real-world applications.

RANDOM NUMBER GENERATOR

A vital characteristic of security is to generate randomness; generally, a security procedure frequently requires the generation of a random number for different objects, for example to generate an asymmetric key, to generate an arbitrary number (nonces) at the time of the authentication phase and to avoid repetitive attacks. It is cryptographically safe when it produces an order that none of the algorithms can predict the next bit of the flow sequence in polynomial time for the preceding bits, with a possibility of greater than half. Mainly, two methods of random number generator are used for cryptographic application: true random number generator (TRNG) and pseudo random number generator (PRNG). TRNC is responsible for measuring a physical disturbance that is expected to be random in nature, like any kind of noise, radiation, etc. Another method PRNG uses is technical algorithms that can produce more random results, but the entire order can be produced if the seed value or key is known. The process of generating random seeds is to exploit a variety of physical phenomena [43]. Randomness sources are restricted only to laptops and desktops, they are not available for resource-constrained devices, hence researchers have tried to design a lightweight PRNG algorithm for such devices [44–46].

SECURE HARDWARE

Most IoT devices can be installed in isolated areas with very low-level security so that an unauthorised user can perform any possible side attacks. Different prevention measures have been suggested in the research, based on different security methods such as encryption [47]. The basic idea of PUF is to generate the unique signature during the manufacture of the chips of each device. Each PUF circuit creates a response corresponding to a given input challenge and these responses are chip-specific due to the very intrinsic hardware [48]. Based on the number of challenge–response pairs supported, PUF can be classified into strong and weak PUF. If PUF can support exponential challenge–response pairs, it is termed as strong PUF [49]. Further research has been suggested for other hardware solutions to prevent various side channel attacks such as implementation of the SIMON algorithm in hardware [50], but all these techniques increase the power consumption of the devices. Because of resource-constrained IoT devices, it could be very difficult to find optimum solutions.

INTRUSION DETECTION SYSTEM

Before preventing the attacks, it is necessary to detect the ongoing attacks. Any antivirus software and traffic controller cannot be used in IoT devices, because they are resource-constrained. For this, researchers have suggested one lightweight intrusion detection system [51], they considered some indicative parameters to detect ongoing attacks like CPU cycle usage, throughput and energy consumption [52].

Table 3.1 presents the summary of various security mechanisms required for efficient operation of IoT services.

MAJOR IOT COMMUNICATION TECHNOLOGIES

A literature survey suggests numerous proposed mechanisms that are specially designed for IoT communication. Some popular communication protocols in the IoT domain are Zigbee, BLE, 6LoWPAN and LoRaWAN. Various IoT communication technologies are explored in the subsections below.

ZIGBEE

Zigbee, like Bluetooth and Wi-Fi, is an open-source short-range wireless protocol developed by Zigbee Alliance. Zigbee standards are specially built for control and

TABLE 3.1

Summary of IoT Security Mechanisms

Security Mechanisms	Description
Encryption [32]	• Technique to ensure confidentiality and integrity • Process of changing the plaintext into ciphertext • Can be easily decrypted (decoded) by a secret key only
Standard encryption mechanism [33–37]	• Used to maintain confidentiality of long messages • Two types of mechanism stream and block cipher mechanism • One of the widely used AES technique
Lightweight cryptography [38–42]	• Cryptographic technique with limited computations and capabilities • Ensure confidentiality, integrity and authenticity • Contains new block, stream cipher and hash functions
Random number generator [43–46]	• Provides randomness in security application • Process to create nonces at the time of authentication • TRNG and PRNG are commonly used random number generators
Secure hardware [47–51]	• Making hardware secure using unique signature • Ensure prevention to any side channel attacks • Improves hardware security
Intrusion detection system [51–53]	• Technique to detect anomalies or any ongoing attacks • Detect attacks with good probability

sensor type networks, the objective is to control and monitor IoT devices. Both Wi-Fi and Bluetooth are not quite suitable for this specific application of wireless communication. It is designed to meet requirement of low-cost communication with a low data rate (20–250 kbps) within short range communication (100–300 m). Zigbee uses advanced encryption standards (AES-128) for data encryption and authentication. In a Zigbee network, mesh topology is used to connect a number of devices; nowadays cluster tree topology is being used to expand the Zigbee network.

There are three types of Zigbee devices. A Zigbee coordinator is the root of the network which performs some important tasks such as channel selection, assigning a unique identifier for the network and allocating a unique address to each device in a network. A Zigbee router acts as an intermediate node between the coordinator and a number of devices. Routers perform functions like routing the traffic between nodes and allowing a number of other devices to join the network. A Zigbee end is responsible for requesting any pending message from a router when it wakes up. There are three frequency channels available for Zigbee networks located in different countries but the coordinator assigns only one channel to the network that has the least interference and that means all devices have to share a single channel to communicate. There are two methods available for channel access: contention-based (devices not synchronised) and contention-free (synchronised devices). Many smart light systems adopt the Zigbee protocol, one of them is the Philips HUE. Zigbee is used for communication by Nest protect smoke alarms and Nest thermostats.

Bluetooth Low Energy

A Bluetooth low-energy (BLE) device, as the name suggests, is a low-power Bluetooth designed for the IoT introduced in 2010 as a part of Bluetooth 4.0 specifications. BLE significantly consumes less power than Bluetooth classic because the radio of BLE devices sleeps between transmission. BLE is ideal for wearable devices like fitness bands, smartwatches and home automation like smart locks and smart lighting. In BLE, there are two types of devices: peripheral devices, also called slaves, are typically constrained devices that need to conserve energy. The central device, also called a master, is typically the device with more processing power and memory. For example, a fitness tracker is a peripheral device and the mobile phone that syncs with a fitness tracker is a central device that processes and displays the data for the user to view. BLE operates in the 2.4 Ghz ISM band which is the same spectrum used by Bluetooth classic. One important point is that a BLE device cannot communicate directly with a Bluetooth classic device.

During the data transmission mode, data are normally shared in the form of fragments to reduce energy consumption. In this manner, peripheral devices can remain on standby mode for a long duration and reactivate periodically to check the channel for any pending messages from the central devices. This reliable communication is possible by a mechanism named stop and wait (S&W) automatic packet retransmission and time-division multiple access (TDMA). The communication through BLE is secured and encrypted by the AES-CCM 128-bit keys algorithm, similar to Zigbee, but the phase of generating a symmetric key during the paring procedure of peripheral and central devices is completely unencrypted [53, 54].

6LoWPAN

6LoWPAN, abbreviated as IPv6, uses low-power wireless over a personal area network. 6LoWPAN is a mesh network in which each node or end has unique IPv6 addresses which helps them to easily communicate with the internet. It supports unicast, multicast and broadcast communication. The architecture of 6LoWPAN consists of three types of devices. A reduced function device (RFD) has fewer sensor capabilities, called RFD or node. A full function device (FFD) has a sensor with higher processing capabilities. A RFD has to first send a data packet to the FFD by hop to hop. The last one is the 6LoWPAN gateway which helps each node to communicate with the internet. The 6LoWPAN protocol stack has an extra layer between the datalink and network layer called the adaption layer. This layer is proposed to make the IPv6 and IEEE standards used in 6LoWPAN compatible with each other. The main function of this layer is header compression and decompression, fragmentation and reassembly of data packets. 6LoWPAN uses AES-128 link layer security for encryption and authentication.

CoAP

Constrained application protocol (CoAP) is a modified internet switching protocol for use with constrained devices and the constrained network of the IoT. The Internet

Engineering Task Force (IETF) designed CoAP to enable limited-resource or constrained devices to join the IoT network with very limited bandwidth and availability. CoAP acts as HTTP protocol for the constrained devices, allowing components like sensing devices and actuators to speak on the IoT. This protocol is meant for continuation where TCP-based protocols like MQTT fail to communicate and exchange information effectively. Methods used by CoAP are equivalent to those employed by HTTP. CoAP depends upon UDP security measures to secure information.

LoRaWAN

LoRaWAN is a link layer protocol developed by the LoRa alliance in 2015 for the operation of low-powered devices [57]. This protocol is placed next to the LoRa physical layer. LoRaWAN is an extended form of star topology that contains gateways, end devices and a network server. In a LoRaWAN network, the end devices can interact through a single channel to a number of gateways. LoRaWAN uses bidirectional communication of channel bandwidth 860/900 MHz ISM band and this communication is always initiated by end-devices. The data transmission rate of this communication ranges from 0.3 to 50 kbps within a range of many kilometres. ALOHA is a protocol to allow to access the channel for communication.

LoRa end-devices are categorised in three different classes: A, B and C. In Class A, after each uplink (ground to satellite) transmission, the end-devices split the downlink into two windows of different sub bands for efficient data reception. The operation mode of Class B allows the end-devices to create extra windows at scheduled intervals to receive the unrequested responses from the network server. Class C devices have a different mode of operation from other classes. Class C devices are connected to the power supply so they always keep the window open for the requests. Over-the-air activation (OTAA) is the predefined procedure through which devices can join the LoRaWAN network.

COMPUTATIONAL INTELLIGENCE-ENABLED CYBERSECURITY FOR IOT

To discover new trends in IoT network cybersecurity, we must learn about computational intelligence (CI)-enabled cybersecurity solutions that should be formed to maintain security in the IoT environment: malware analysis, data security, security architecture, security incident management and response, intrusion detection, threats and incident analysis and more. CI-based algorithms include reasoning and decision-making techniques, data mining concepts, image and language processing, biologically inspired algorithms, and artificial immune systems. This new type pf algorithms are used to design cybersecurity applications. CI-based techniques can provide solutions to cybersecurity problems that might be difficult to resolve by using conventional computational algorithms.

The CI-based technique can improve existing cybersecurity systems by following three levels: prevention and protection, it can be observed that conventional security solutions such as IDS are still struggling to cope with security threats. New emerging cyberthreats like PowerShell attacks and fileless attacks are not easy to detect. CI-based security solutions can easily catch malicious source code and fix it in an appropriate and computerised way. CI-enabled techniques can be deployed in cyber defence tools that can learn from emerging threats, characterised data and the capability to develop security planning on its own. CI-based security solutions can help to meet the need of protecting data and assets in IoT systems. The second level is to identify and detect threats, standard machine-learning algorithms such as unsupervised learning and deep learning are being used in detecting the threat pattern in provided datasets and these analyses can be further used to determine breaches in cyber behavior. Similarly, CI algorithms can also be taken into account in threat analysis. Several self-training CI algorithms can be used to detect threats present in a system that cannot be figured out by humans from normal behavior. Symantec, DarkTrace, Cylance, etc. are well-known security solution providers using CI-based techniques to provide optimised security. The last level is response. The IoT network periodically produces a huge amount of data that should go through CI-based security solutions so that solutions can detect breaches or threats rapidly and compute an optimised response. CI-enabled cyberse-curity tools can conduct operations like the detection of threats, searching and security analysis automatically without any human intervention. CI-based tools have the capa-bility to learn from potential threats and to develop a response plan or strategy rapidly; this self-healing process reduces human intervention. The basic mechanism behind anomaly detection using computational intelligence is depicted in Figure 3.3.

CONCLUSION

This study concludes that a lack of security in the IoT network allows hackers and intruders to access sensitive data. The chapter presents cybersecurity issues and solu-tions related to the five layers of IoT architecture. It highlights the security concerns and issues associated with each layer in order to maintain privacy and enhance trust. Considering the resource-constrained nature of IoT devices, this work presents a detailed discussion on various types of associated communication technologies.

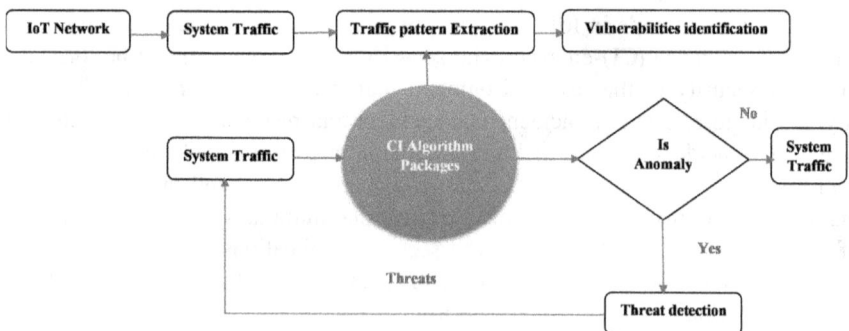

FIGURE 3.3 Computational intelligence for anomaly detection.

Further, the most prominent security techniques for IoT services like standard encryption mechanism, cryptography, secure hardware and intrusion detection system (IDS) are discussed. But we have also seen that these mechanisms cannot be used directly on IoT devices due their constrained nature. Therefore, numerous researchers have suggested modified versions of these mechanisms like lightweight cryptography, lightweight intrusion detection systems, etc., that can operate on such devices, using secure hardware and a random number generator for the authenticity of devices. This chapter presents the major security techniques related to communication protocols used in the IoT like Zigbee, 6LoWPAN, etc. Finally, the chapter outlines the need for intelligent systems in order to enhance the overall security approaches to achieve an optimised result.

REFERENCES

1. Lee, I. (2019). The Internet of things for enterprises: An ecosystem, architecture, and IoT service business model. *Internet of Things and Cyber-Physical Systems*, 7, 100078.
2. Atzori, L., Iera, A. & Morabito, G. (2010). The internet of things: A survey. *Computer Network*, 54(15), 2787–2805.
3. Irdeto. (2019). New 2019 global survey: IoT-focused cyberattacks are the new normal. https://resources.irdeto.com/global-connected-industries-cybersecurity-survey /new-2019-global-survey-iot-focused cyberattacks-are-the-new-normal (accessed on September 17, 2020).
4. Saxena, S., Bhushan, B. & Ahad, M. A. (2021). Blockchain based solutions to secure Iot: Background, integration trends and a way forward. *Journal of Network and Computer Applications*, 103050. https://doi.org/10.1016/j.jnca.2021.103050
5. Sharma, N., Kaushik, I., Agarwal, V. K., Bhushan, B. & Khamparia, A. (2021). Attacks and security measures in wireless sensor network. *Intelligent Data Analytics for Terror. Threat Prediction*, 237–268. https://doi.org/10.1002/9781119711629.ch12
6. Warner, M. R. (1691). Internet of Things cybersecurity improvement act of 2017, *S. 1691, 115th US Congress*, September 2017.
7. Saied, Y. B. (2013, June). *Collaborative security for the Internet of Things*. Ph.D. Dissertation, Institut National des Télécommunications.
8. Goyal, S., Sharma, N., Kaushik, I. & Bhushan, B. (2021). Blockchain as a solution for security attacks in named data networking of things. *Security and Privacy Issues in IoT Devices and Sensor Networks*, 211–243. https://doi.org/10.1016/b978-0-12-821255 -4.00010-9
9. Arora, A., Kaur, A., Bhushan, B. & Saini, H. (2019). Security concerns and future trends of Internet of things. In *2019 2nd International Conference on Intelligent Computing, Instrumentation and Control Technologies (ICICICT)*. https://doi.org/10 .1109/icicict46008.2019.8993222
10. Mollah, M. B., Azad, M. A. & Vasilakos, A. (2017). Security and privacy challenges in mobile cloud computing: Survey and way ahead. *Journal of Network and Computer Applications*, 84, 38–54. https://doi.org/10.1016/j.jnca.2017.02.001
11. Gao, Y., Ranasinghe, D. C., Al-Sarawi, S. F., Kavehei, O., & Abbott, D. (2016). Emerging physical unclonable functions with nanotechnology. *IEEE Access*, 4, 61–80. https://doi.org/10.1109/access.2015.2503432.
12. Kulseng, L., Yu, Z., Wei, Y. & Guan, Y. (2010). Lightweight mutual authentication and ownership transfer for RFID systems. In *Proceedings of the 2010 IEEE INFOCOM*, San Diego, CA, March 15–19, 2010 (pp. 1–5).

13. Xu, H., Ding, J., Li, P., Zhu, F. & Wang, R. (2018). A lightweight RFID mutual authentication protocol based on physical unclonable function. *Sensors, 18*(3), 760.

14. Goel, A. K., Rose, A., Gaur, J. & Bhushan, B. (2019). Attacks, countermeasures and security paradigms in IoT. In *2019 2nd International Conference on Intelligent Computing, Instrumentation and Control Technologies (ICICICT)*. https://doi.org/10.1109/icicict46008.2019.8993338

15. Li, J., Zhao, Z., Li, R. & Zhang, H. (2019). AI-based two-stage intrusion detection for software defined IoT networks. *IEEE Internet of Things Journal, 6*(2), 2093–2102.

16. Pacheco, J., Benitez, V. & Félix, L. (2019). Anomaly behavior analysis for IoT network nodes. In *Proceedings of the 3rd International Conference on Future Networks and Distributed Systems*, Paris, France, 1–2 July 2019 (pp. 1–6).

17. Singh, R. V., Bhushan, B. & Tyagi, A. (2021). Deep learning framework for cybersecurity: Framework, applications, and future research trends. *Advances in Intelligent Systems and Computing*, 837–847. https://doi.org/10.1007/978-981-33-4367-2_80

18. Soni, S. & Bhushan, B. (2019). Use of machine learning algorithms for designing efficient cyber security solutions. In *2019 2nd International Conference on Intelligent Computing, Instrumentation and Control Technologies (ICICICT)*. https://doi.org/10.1109/icicict46008.2019.8993253

19. Pajouh, H. H., Javidan, R., Khayami, R., Dehghantanha, A., & Choo, K.-K. R. (2019). A two-layer dimension reduction and two-tier classification model for anomaly-based intrusion detection in IOT backbone networks. *IEEE Transactions on Emerging Topics in Computing, 7*(2), 314–323. https://doi.org/10.1109/tetc.2016.2633228.

20. Bhushan, B., Sahoo, C., Sinha, P. & Khamparia, A. (2020). Unification of Blockchain and Internet of Things (BIoT): Requirements, working model, challenges and future directions. *Wireless Networks*. https://doi.org/10.1007/s11276-020-02445-6

21. Sharma, T., Satija, S. & Bhushan, B. (2019). Unifying Blockchain and IoT: Security requirements, challenges, applications and future trends. In *2019 International Conference on Computing, Communication, and Intelligent Systems (ICCCIS)*. https://doi.org/10.1109/icccis48478.2019.8974552

22. Sinha, P., Rai, A. K. & Bhushan, B. (2019). Information Security threats and attacks with conceivable counteraction. In *2019 2nd International Conference on Intelligent Computing, Instrumentation and Control Technologies (ICICICT)*. https://doi.org/10.1109/icicict46008.2019.8993384

23. Chowdhury, A. & Raut, S. A. (2018). A survey study on Internet of Things resource management. *Journal of Network and Computer Applications, 120*, 42–60. https://doi.org/10.1016/j.jnca.2018.07.007

24. Tran-Dang, H. & Kim, D. (2018). An information framework for Internet of things services in physical Internet. *IEEE Access, 6*, 43967–43977. https://doi.org/10.1109/access.2018.2864310

25. Choo, K. R., Gritzalis, S. & Park, J. H. (2018). Cryptographic solutions for industrial Internet-of-things: Research challenges and opportunities. *IEEE Transactions on Industrial Informatics, 14*(8), 3567–3569. https://doi.org/10.1109/tii.2018.2841049

26. Zarpelão, B. B., Miani, R. S., Kawakani, C. T. & Alvarenga, S. C. D. (2017). A survey of intrusion detection in Internet of Things. *Journal of Network and Computer Applications, 84*, 25–37. https://doi.org/10.1016/j.jnca.2017.02.009

27. Hamad, S. A., Sheng, Q. Z., Zhang, W. E. & Nepal, S. (2020). Realizing an Internet of secure things: A survey on issues and enabling technologies. *IEEE Communications Surveys and Tutorials*, 1–1. https://doi.org/10.1109/comst.2020.2976075

28. Butun, I., Osterberg, P. & Song, H. (2020). Security of the Internet of things: Vulnerabilities, attacks, and countermeasures. *IEEE Communications Surveys and Tutorials, 22*(1), 616–644. https://doi.org/10.1109/comst.2019.2953364

29. Lao, L., Li, Z., Hou, S., Xiao, B., Guo, S. & Yang, Y. (2020). A survey of IoT applications in blockchain systems. *ACM Computing Surveys, 53*(1), 1–32. https://doi.org/10.1145/3372136

30. Maharaja, R., Iyer, P. & Ye, Z. (2019). A hybrid fog-cloud approach for securing the Internet of Things. *Cluster Computing.* https://doi.org/10.1007/s10586-019-02935-z

31. Kumar, S. A., Vealey, T., & Srivastava, H. (2016). Security in Internet of things: Challenges, solutions and future directions. In *2016 49th Hawaii International Conference on System Sciences (HICSS).* https://doi.org/10.1109/hicss.2016.714

32. Siddiqui, F., Beley, J., Zeadally, S. & Braught, G. (2019). Secure and lightweight communication in heterogeneous IoT environments. *Internet of Things,* 100093. https://doi.org/10.1016/j.iot.2019.100093

33. Rivest, R. L., Shamir, A. & Adleman, L. M. (1983). Cryptographic communications system and method. U.S. Patent US4, 405, 829A, September 20, 1983.

34. McEliece, R. J. (1978). A public-key cryptosystem based on algebraic coding theory. *Deep Space Network Progress Report, 44,* 114–116.

35. Elgamal, T. (1985). A public key cryptosystem and a signature scheme based on discrete logarithms. *IEEE Transactions on Information Theory, 31*(4), 469–472.

36. Stallings, W. (2014). *Cryptography and network security principle and practice* (6th ed.). Pearson.

37. Information Technology. (2011). Security techniques – Message authentication codes – Part 1: mechanisms using a block cipher, ISO/IEC, 9797–1:2011, ISO/IEC std., March 2011.

38. Information Technology. (2012). Security techniques – Lightweight cryptography, ISO/IEC 29192–2012, ISO/IEC std., January 2012.

39. Weber, R. H. (2010). Internet of Things – New security and privacy challenges. *Computer Law and Security Review, 26*(1), 23–30.

40. Fremantle, P. & Scott, P. (2017). A survey of secure middleware for the Internet of Things. *PeerJ Computer Science, 3,* e114.

41. Guo, J., Peyrin, T. & Poschmann, A. (2011). The PHOTON family of lightweight hash functions. In *Proceedings of the Annual Cryptology Conference, 6841,* 222–239.

42. Bogdanov, A., Kneževic, M., Leander, G., Toz, D., Varıcı, K. & Verbauwhedé, I. (2011). SPONGENT: A lightweight hash function. In *Proceedings of the International Workshop on Cryptographic Hardware and Embedded Systems, 6917,* 312–325.

43. Eastlake, D., Schiller, J. & Crocker, S. (2005). Randomness requirements for security. *RFC, 4086* (June).

44. Mandal, K., Fan, X. & Gong, G. (2016). Design and implementation of warbler family of lightweight pseudorandom number generators for smart devices. *ACM Transactions on Embedded Computing Systems, 15*(1), 1.

45. Orúe López, A. B., Hernández Encinas, L., Martín Muñoz, A. & Montoya Vitini, F. (2017). A lightweight pseudorandom number generator for securing the Internet of Things. *IEEE Access, 5,* 800–827.

46. Bakiri, M., Guyeux, C., Couchot, J., Marangio, L. & Galatolo, S. (2018). A hardware and secure pseudorandom generator for constrained devices. *IEEE Transactions on Industrial Informatics, 14*(8), 3754–3765.

47. Halak, B., Zwolinski, M. & Mispan, M. S. (2016). Overview of PUF-based hardware security solutions for the Internet of Things. In *Proceedings of the IEEE 59th International Midwest Symposium on Circuits and Systems,* October 2016 (pp. 1–4).

48. Roel, M. (2012). *Physically unclonable functions: Constructions, properties and applications.* Springer.

49. Suh, G. E. & Devadas, S. (2007, June). Physical unclonable functions for device authentication and secret key generation. In *Proceedings of the 44th ACM/IEEE Design Automation Conference* (pp. 9–14).

50. Begam, S., Vimala, J., Selvachandran, G., Ngan, T. T. & Sharma, R. (2020). Similarity measure of lattice ordered multi-fuzzy soft sets based on set theoretic approach and its application in decision making. *Mathematics, 8*, 1255.

51. Vo, T., Sharma, R., Kumar, R., Son, L. H., Pham, B. T., Tien, B. D., Priyadarshini, I., Sarkar, M. & Le, T. (2020). Crime rate detection using social media of different crime locations and Twitter part-of-speech tagger with Brown clustering. *Journal of Intelligent & Fuzzy Systems, 38*(4), 4287–4299.

52. Nguyen, P. T., Ha, D. H., Avand, M., Jaafari, A., Nguyen, H. D., Al-Ansari, N., Van Phong, T., Sharma, R., Kumar, R., Le, H. V., Ho, L. S., Prakash, I. & Pham, B. T. (2020). Soft computing ensemble models based on logistic regression for groundwater potential mapping. *Applied Sciences, 10*, 2469.

53. Jha, S., Kumar, R., Hoang Son, L., Abdel-Basset, M., Priyadarshini, I., Sharma, R. & Viet Long, H. (2019). Deep learning approach for software maintainability metrics prediction. *IEEE Access, 7*, 61840–61855.

54. Sharma, R., Kumar, R., Sharma, D. K., Priyadarshini, I., Pham, B. T., Bui, D. T. & Rai, S. (2019). Inferring air pollution from air quality index by different geographical areas: Case study in India. *Air Quality, Atmosphere and Health, 12*, 1347–1357.

4 Cyberattacks and Risk Management Strategy in Internet of Things Architecture

Saurabh Bhatt and Bharat Bhushan

CONTENTS

DOI: 10.1201/9781003097518-4

INTRODUCTION

Since the birth of the Internet of Things (IoT) it has been growing and now it plays an important role in our daily lives, having a role in our workplace and even in our homes. As an emerging technology it has proved to revolutionise the global network consisting of data, information, intelligent and smart devices and humans [1, 2]. The IoT has evolved over time and is still evolving so it is hard to define it, but it can be best defined as the system of portable devices connected by a wireless network without human interference for data collection and transmission. The IoT can be divided into four elements, these are things, analytics, processes and people. Things consist of the physical gadgets or the interconnected objects. The analytics element processes and analyses individual data streams using algorithms. The processes are the transmission of data or information at the correct time. People are the consumers receiving valuable information about their things or receiving related services [3]. Connected equipment at the industrial level can be used to improve process operational efficiency, solve engineering problems faster or predict and prevent equipment failures [4, 5]. This technology offers us the opportunity to ease our lives by understanding and improving our daily needs. IoT devices are considered to be smart devices as they can connect with each other individually. Every IoT device has its own functions and features which can be used individually or in conjunction with other IoT devices or non-IoT devices. These features are data capabilities, support skills, transducer capabilities and interface capabilities. Digital capabilities are used to conduct digital calculations which include processing and storing. Supporting capabilities involve functions like cybersecurity and device management. Transducer capabilities are used for direct interaction with physical objects. Lastly, IoT gadgets use interface abilities to interact with other IoT devices [6].

In addition to these benefits, being connected to the internet also means that we are connected to potential cyberthreats. When the refrigerator is connected to the internet, it becomes a device that can be used by cybercriminals, like a laptop or cell phone. The more devices we connect in a network, the more control and access we give to attackers. As the IoT market expands, the number of potential risks that threaten device performance and security and the integrity of IoT data also increases [7]. Attackers are always looking for new ways to compromise systems and gain access to devices and data systems. From smart medical devices to the navigation systems of internet-connected vehicles, any IoT system can become a target for hackers.

As the IoT has become so common, our surroundings are becoming connected with more IoT devices, hence, making a proper IoT environment around us. Because of this rapid growth many security risks have also been discovered in IoT devices which can affect our lives negatively [8]. From building access control to video recordings of private environments, these are the main source from where attackers can gain sensitive information or control over the device or network. There are many cyberattacks that attackers can use to bypass IoT security systems as they have very low-level security measures. Many mass attacks have also been reported in the past few years; one of the most popular was the Mirai botnet [9, 10].

The main aim of this chapter is to provide an updated visualisation of the current cybersecurity situation in the IoT field. The major contributions of this work are as follows:

- An in-depth review of cybersecurity-oriented IoT architecture.
- An analysis of the major types of cyberthreats and risks affecting IoT devices.
- An evaluation of these cyberthreats and their plausible countermeasures.
- An introduction to artificial intelligence and its correlation with IoT cybersecurity.
- A comparative study of quantum-resistant cybersecurity technologies.

The chapter is organised as follows. The next section describes the cybersecurity-oriented IoT architecture. Here all the four layers of the architecture are discussed, starting with the sensing layer, then the network layer, after that middleware layer and finally the application layer. The following section describes various cyberattacks that are carried out on IoT devices, like MITM attacks, physical attacks, DoS attacks and more. Then there is a section dedicated to the two types of approaches in cybersecurity risk management. Next is a presentation of the co-relation of cybersecurity, IoT and AI. This is followed by a section which explores quantum-resistant cybersecurity technologies, leading to a conclusion section.

CYBERSECURITY-ORIENTED IoT ARCHITECTURE

The term IoT stands for Internet of Things. It holds within it everything connected to the internet. In IoT systems, things are connected via smart devices or gadgets such as smartphones, RFID readers and tags, etc. There is a four-layered cybersecurity-oriented architecture for IoT [11].

Sensing Layer

This layer contains data sensors. Here the networks can detect the data, collect the data, process the data and finally transmit data to the whole network. There are basically four cybersecurity problems, these are: the dynamic nature of IoT devices, the disclosure of sensor nodes in IoT gadgets, the strength of wireless signals and storage and memory, computation and communication constraints. This layer has three techniques to defend the IoT system: the access control technique, lightweight encryption technique and nodes authentication technique. In this layer the attacks are basically focused on confidentiality [12]. Some of the attacks used for this layer are side-channel attacks [13], node capture attacks [14], replay attacks [15], malicious data attacks [16], timing attacks [17], etc. In a replay attack valid data channelling is maliciously recurrent or delayed. It is also known as a playback attack. In a timing attack, a cryptosystem is compromised by examining the time taken to execute cryptographic algorithms. In a node capture attack, as the name suggests the attacker gains access to take over the nodes and then captures useful information and data. In continuation of a node capture attack the attacker can use a malicious data attack by sending the malicious input to the layer by adding a new node. In a side-channel attack data is extracted from a system by analysis of physical parameters such as electromagnetic emission, execution time, etc.

NETWORK LAYER

The function of a network layer is to route and transmit the data to various IoT hubs and gadgets over the internet. Technologies such as, 3G/4G, Bluetooth and Wi-Fi are used here for transmission of data. The network gateways work as intermediaries between different IoT nodes by filtering, transmitting and aggregating data through sensors. Cybersecurity issues here include confidentiality, privacy and compatibility. The interconnection in IoT devices is operated by wireless or wired mechanisms among various smart objects or devices. As the whole system is so implanted in this layer, attackers have good odds of revealing criminal activities [12]. A man-in-the-middle attack has proven to be very effective for this layer. Apart from this, some basic direct attacks on this layer used by attackers are spoofing [18], altering, routing information, sybil [19], wormholes [20], etc. Spoofing, altering a replay and target data exchange, is used mutually, generating false messages and creating routing loops between the nodes are achieved by attackers. Sybil attacks are found in peer-to-peer networks. Here a node in the network functions with many identities active at the same time. It steals data by reducing integrity, spreading malware and resource utilisation within the IoT. A wormhole attack is a grave attack, it basically keeps listening to the network and stealing wireless information.

MIDDLEWARE LAYER

The functions of the middleware layer include collecting and filtering the acknowledged data from hardware devices then execution of information discovery and handing access control to the devices for applications. It is based on the principle of service-oriented architecture. Here, attackers focus on integrity, authenticity and confidentiality. Some of the attack types used by attackers for this layer are malicious insider [21], underlying infrastructure, third-party relationship and virtualisation. Malicious insider attacks are also known as turncloak attacks. Here an internal attacker purposely modifies and extracts data inside the network. Underlying the attack is a platform-as-a-service-based attack. In platform-as-a-service, the aim of the developer is to keep the IoT secure. Third-party relationship attacks, as the name suggests, are induced by third-party components. Hyperjacking is a type of virtualisation attack. In this, the attacker makes a malicious command over the hypervisor which produces the virtual environment within a virtual machine host.

APPLICATION LAYER

In this layer various application services are delivered. These services are provided to various applications and users in IoT-based systems through the middleware

TABLE 4.1
Cyber-Oriented IoT Architecture

Layer	Attacks	Target sector
Sensing layer	Side-channel attacks, node capture attacks, replay attacks, malicious data attacks, timing attacks	Confidentiality
Network layer	Spoofing, altering, routing information, sybil, wormholes	Confidentiality, privacy and compatibility
Middleware layer	Malicious insider, third-party relationship and virtualisation, underlying infrastructure	Integrity, authenticity and confidentiality
Application layer	Phishing attack, worms, virus, spyware, malicious scripts, Trojan horse	Data privacy and identity authentication

layer. This layer delves into all the system functionalities for final users. The focus targets of attackers in this layer are data privacy and identity authentication. Data access permissions, data protection and recovery, software vulnerabilities and the ability to deal with mass data are some common issues faced in this layer. Some of the attack types used by attackers for this layer are phishing attacks [22], worms [23], viruses [24], spyware [25], malicious scripts [26], trojan horse [27] and unauthorised access. All these attacks can be used mutually. In phishing, the attacker masquerades as a trustworthy entity, tricks the user into opening an email, a text message or a website. Then the attacker gains useful data from the user, like login credentials or credit card information. Finally, the attacker uses that information to inject malware into the system using a virus, trojan horse or spyware and accessing confidential data [12].

From a cybersecurity perspective, the IoT structure is a four-layer structure and is depicted in Table 4.1.

ATTACKS ON IoT DEVICES

Because of inadequate security in IoT gadgets cyberattackers have found various methods for attacking IoT gadgets from various diverse attack surfaces. According to many studies, IoT gadgets have drastically increased in number around us and so the attacks on them are also drastically increased. Combining the increasing attack speed with the increased number of devices, we can see why appliance managers are more worried than ever. Due to the inadequate security controls, IoT gadgets lack the functionality to detect vulnerabilities [28]. Hence, the risk of cyberattacks is very real. A few of the attacks or techniques are discussed below.

INITIAL RECONNAISSANCE

Before attacking any device, the cyberattackers first do a footprinting of the target, in this case the target is the IoT devices. It helps them to understand the device's security posture. By properly researching the target, the attacker can gain data and information on the target. The attacker may even buy the IoT device from the market. Then they try to reverse engineer the device so that they can do their test attack on it. The test attack gives the idea of the outcome of the attack and the likelihood of attacking the device. The attacker can also open up the IoT device and examine the internal hardware for software information and manipulate the microcontroller to identify confidential information [28]. Any information that can be gained is important as it could identify any entry points into the networks. To prevent reverse engineering the IoT devices should have hardware security installed. Device verification can also be performed using hardware security so that the device can assure the sever it is linked to that and is not fraudulent [29].

PHYSICAL ATTACKS

In a physical attack, the attacker gains physical access to a physical asset. Here the physical component of the targeted device is used to gain advantage by the attacker in few ways. There are many types of physical attacks like outage attacks, in which the network to which the devices are connected is turned off and they cease the function or some physical damage in which the device is damaged so that it stops functioning properly and also by injecting malicious code, the attacker uses jammers to jam or manipulate the signals generated or used by device. As stated above, the attackers may also buy a physical copy of the device and then perform some reverse engineering on it by opening the device up [28]. Permanent denial of service (DoS) attacks can also be used as a physical attack. Physical attacks are often used to identify the vulnerabilities of the device.

MAN-IN-THE-MIDDLE ATTACK

A man-in-the-middle (MITM) attack is an attack where the attacker positions themself in between the user and an application. The attacker does this either to eavesdrop or to impersonate one of the parties. Login credentials, bank account details or credit card details some of the details which the attacker tries to steal. With the IoT, MITM is one of the most well-known attacks. This kind of attack is widely used in the network layer of the IoT architecture. In the IoT, the attacker positions themself between many different connections such as a server and a client. As IoT devices do not have standard implementations to withstand any types of attacks they are more vulnerable to MITM attacks. There are basically two common methods of MITM attacks: direct connection and cloud polling. Attackers can either redirect network traffic by altering the domain name system (DNS) or they can intercept HTTPS traffic by self-signed certificates. Probably the best way to avoid MITM attacks is by using a strong encryption method between the client and server. Digital certificates can be used to prove identity and can be installed on every device. After all, identity is the major factor causing MITM attacks. Another method to prevent this attack is by using a secured and encrypted VPN [30, 31].

RANSOMWARE

A malware which encrypts the target's resource then demands the target to pay the amount of ransom for accessing the encrypted resource is called ransomware. There are two types of ransomware attacks, locked ransomware and crypto ransomware. Mostly, all of these attacks are carried out via phishing techniques, where the attacker basically sends an attachment or a link which, when opened, causes malware to attack the device. In crypto ransomware, the malware encrypts some of the important files or even the whole hard disk. On the other hand, locked ransomware encrypts or locks the whole computer of the target and then demands a ransom for the decryption key. Usually, these ransomware attacks can interfere with our day-to-day operations or work and mostly these ransom demands are high [32]. Major targets of these types of attacks are law firms, the healthcare sector and educational institutions. As the majority of IoT devices do not have any display system, the attackers have to work more and discover a way to communicate with the target by getting their email nor even hacking a certain application that can control the device. These attacks generally happen in the application layer of the IoT.

BRUTE FORCE ATTACKS

Brute force attacks are used to gain the sensitive credentials of the target by conducting a series of guess-based trials of the credentials. Attackers try every possible combination to make an accurate guess. As the name suggests brute force means when an attacker tries to force themself into the target's personal account by using excessive force. On the basis of the complexity of the password or key it can take a huge amount of time to crack it. These attacks also come under the application layer. Brute force attacks are divided into five types, these are credential stuffing, simple brute force, hybrid brute force, dictionary attacks and reverse brute force attacks. In credential stuffing the attackers use the known username and password combination which worked for a particular website in the hope that the target must have used the same combination in other platforms as well. In simple brute force, attackers try to guess the credentials logically without any software assistance. Dictionary attacks play an important role in brute force attacks as these consist of dictionaries of possible passwords, words or special characters. These dictionary attacks are an important tool in brute force attacks. Hybrid brute force attacks are the mixture of simple brute force attacks and dictionary attacks. Reverse brute force attacks are the opposite of simple brute force attacks, instead of finding the password, usernames are searched [33].

BOTNETS

A botnet, short for robot network, is number of computers connected with each other and infected by malware under the control of a single computer, a bot-herder. Botnets are used for various purposes like distributed DoS attacks, data theft and many others. Apart from MITM, botnet attacks are also common on IoT devices. IoT devices become easy targets for bot-herders because the security development lifecycle of an IoT device is often rushed or sometimes even bypassed due to the deadlines of the market or the cost of hardware. As many IoT devices use a stripped-down version of Linux as the operating system (OS) it makes it easy to compile malware with their OS. IoT devices like CCTV cameras have full access to the internet and they are not subject to bandwidth limitation. A lot of hardware and software is reused in the development of IoT devices, this shows a lot of security-critical components shared between devices [34]. A botnet virus resides in the memory of the device not in the hard disk. So once the device is restarted or rebooted the virus is gone. The best and first way to keep away from botnets is to keep updating the IoT device. Using physical authentication keys is also a good way of getting out of trouble.

DENIAL-OF-SERVICE (DoS) ATTACK

A denial-of-service or DoS attack is an attack in which the resources of a website are tied up so that the users cannot access the website. These attacks basically flood the web services or crash them. IoT devices are easily damaged by DoS attacks. Permanent denial-of-service (PDoS) attacks have a major effect on IoT devices making them completely useless. As discussed above, a lot of hardware is reused in the development of IoT devices, most of that from cheap hardware components [35]. This hardware and firmware usually contain various security vulnerabilities. The majority of IoT devices are almost always turned on and always connected to internet and are interconnected with each other making them easily visible on the internet, hence they are constantly being pounded with everything the internet has to offer them, including malware payloads making them more vulnerable to DoS attacks. In IoT devices the DoS attack bricks or destroys the firmware making the device useless. Uploading a corrupted BIOS (Basic Input Output System) to the device can also carry a PDoS attack. These attacks can also be carried out physically by using a malware-infected USB stick. The best way to keep PDoS and DoS attacks at bay from IoT devices is by keeping the devices upgraded and patched. Table 4.2 provides the summary of the listed attacks.

CYBERSECURITY RISK MANAGEMENT

Similar to real-world risk management, in cybersecurity risk management the risks and vulnerabilities are identified, and actions and solutions are used for safety [36]. Many studies have classified the cybersecurity risk management into two parts, qualitative and quantitative approaches.

QUALITATIVE APPROACH

It prioritises in the recognised project risks by using a predefined evaluation scale. Here the risks are scored based on their possibility of happening and their impact

TABLE 4.2

Attacks on IoT Devices

Type of Attack	Description
Initial reconnaissance	These are just the initial steps an attacker takes before attacking their target; attackers basically try to footprint their target and study them to find their vulnerabilities.
Physical attack	Here the hardware of the target is used for gaining an advantage. Outage attacks are commonly used here or even physical attacks on the target so that it starts malfunctioning.
MITM	In MITM attacks the attacker mostly acts as a conduit between the app and the user. Sometimes the attacker may also just position themselves in between for eavesdropping purposes.
Ransomware	This is malware which encrypts or locks some of important files or even the whole system then the attacker demands a ransom in exchange for the decryption key.
Brute force	One of the oldest methods, brute force attacks are attacks used to steal personal credentials with excessive force, with the attacker forcing themself inside personal accounts.
Botnet	These are networks of interconnected computers all infected with malware under the control of a single computer.
DoS attack	Here the attacker makes the targeted resource inaccessible by flooding it with loads of traffic over the internet.

on the organisation. The qualitative approach comes under the empirical analysis of gaps in cyber risk impact assessment. There are many competing risk management frameworks in the world. Some frameworks like Operationally Critical Threat, Asset and Vulnerability Evaluation (OCTAVE) offer diverse qualitative methods of risk management [36]. OCTAVE identifies the key assets of the organisation then considers the threats to the assets and finally analyses their vulnerabilities and impacts, after all, this OCTAVE develops some security policies and priorities to reduce the risk to the asset. Another well-known cybersecurity framework is The NIST Cybersecurity Framework. The NIST Cybersecurity Framework has been divided into three major parts: tiers, profile and core. The core part contains a range of activities, insights and references related to cybersecurity approach. The profile comprises a list of results that the organisation had selected from categories based on their need and risk assessments.

The International Organization for Standardization (ISO) has also developed a set of standards regarding the information of security risks that is ISO/IEC 27005. This set of standards does not enlist any specific risk management methods. They let the organisations make their own approach towards risk management. Exostar systems are used for identifying cyber risks including supply chain cyberattacks. This system has five informative points for assessing the supply chain aspect of the risk. By eliminating the supply chain risk, the current general state of cyber maturity can be verified by Capability Maturity Model Integration (CMMI). The CMMI has similar approach to risk management as the OCTAVE. One of the distinctive features of the cyber maturity platform is the risk-based roadmap which consists of the prioritised list of customised actions based on most relevant risk for the organisation. Other than these frameworks, there are even some methodologies for measuring cyber risk. A good example of one of these methodologies is Threat Assessment and Remediation Analysis (TARA), it applies a threat matrix as a qualitative approach [37].

QUANTITATIVE APPROACH

This approach narrows down the scope of studies to a security assessment. In this assessment numerical ratings are assigned; here, further analysis of the highest priority risks is done. The numerical ratings are later used to form a probabilistic analysis of the project. The qualitative risk analysis provides a quantitative approach for making decisions when there is uncertainty. Probabilistic graphical models like the Bayesian decision network (BDN) were applied to frameworks for network security risk management. These security assessment frameworks consist of various processes like risk assessment, risk mitigation and risk validation and monitoring. Information about vulnerabilities, risk-reducing counter measures and the effect of counter measures on vulnerabilities are fed to the BDN model for managing security risks [37]. For better accuracy of the risk assessment process, the probability of exploitation of vulnerabilities and the impact of exploitation of vulnerabilities resources are calculated using intrinsic, temporal and environmental factors.

AVARCIBER, another risk assessment framework, stands for assets, vulnerabilities, threats, risks of cybersecurity. ISO 27005 has been used as a reference for AVARCIBER, just extending few of the parameters of the set of standards developed by ISO 27005. This framework has many procedural steps involved in the implementation of the framework. First of all, the risk assessment is launched after that the assets are identified and accessed. In this step the organisation identifies the security dimensions, which is basically CIA and stands for confidentiality, integrity and availability, of the assets and measures the impact of the asset. After this, in the third step, the damage level of the vulnerability of the asset is determined. Then the cybersecurity threats are identified. Here the vulnerability of the asset that may materialise for a cyberthreat is determined [36, 37]. The measurement of the risk is then conducted. Finally, some countermeasures are performed which can mitigate or eliminate the potential threats.

ARTIFICIAL INTELLIGENCE (AI) IN CYBERSECURITY

As majority of cybersecurity specialists have already accepted AI to be the future of our world, they have also favored AI in regards to cybersecurity. AI can take just few milliseconds to react to cyberattacks so is called a cybersecurity expert and is being used in cybersecurity to look into the characteristics of attacks [38].

MACHINE LEARNING

Machine learning (ML) is an approach to AI. Here, the model learns from the experience. ML is divided into three parts, reinforcement learning, unsupervised learning and supervised learning. In supervised learning the machine is trained with the used of labelled data. On the basis of the previous experience the machine learns by itself. In unsupervised learning there is no labelled data, the algorithm acts on its own on the given data [38]. Here, the algorithm groups similar data by itself. In reinforcement learning the machine is made to take suitable action to maximise the reward in particular situations. ML used for network security uses regression, classification and clustering. Clustering is used for forensic analysis, regression is used for predicting network packet parameters and classification is used for identifying classes of network. One of the well-known ML algorithms, naïve Bayes, is used in cybersecurity to classify data based on Bayesian theorem. Clark et al. [39] proposed that a robotic vehicle based on ML policy performed well when malicious attacks were tested on it. In another work, Xin et al. [40] suggested that use of ML and DL (Deep Learning) methods for intrusion detection on networks have shown positive response on the major side. Similarly, Azwar et al. [41] has shown the use of ML and data mining for intrusion detection in secure networks using the dataset CICIDS2017 for better accuracy. Zhang et al. [42] introduced the brute force black box method for the evaluation of the strength of ML classifiers in cybersecurity using many examples. In other work, Li et al. [43] devised an incentive framework based on ML for malware detection by a domain generation algorithm. Furthermore, Ortiz et al. [44] proposed a brilliant study how ML can help us in detecting and help us against phishing attacks.

DECISION TREES

A decision tree is a tool used for classification and prediction. It uses a tree-like model for decisions and the possible outcomes of those decisions. It consists of a root node, with the root decision followed by further nodes. It is a flowchart-like structure. Decision trees are used in cybersecurity for detecting various attacks or the paths of attacks. Usually, they are used to detect DoS attacks. When the data flow rate is low, but the traffic is taking a much longer time it is a possibility that it is a DoS attack. Command-injecting attacks are also detected using decision trees in driverless cars. There are some decision tree applications related to cybersecurity in the market too. For example, SecurITree – this tree helps the users to determine the likely paths of attacks. There is another approach using decision trees in which the tree looks for a set of attack features or characteristics at each repetition by maximising a certain score that indicates the quality of the classification, this approach is known as the rule-learning technique [38].

K-NEAREST NEIGHBORS

K-nearest neighbors, also known as kNN, is an approach for data classification which tries to determine to which group a data point belongs. This technique is learning from the data samples and form classes by examining Euclidean distance among new data fragments and already classified data fragments to choose into which class to place the new element. This is widely used for intrusion detection as it can easily learn from new traffic patterns. One of the methods in kNN is the text categorisation method; in this, the text documentations are grouped together based on the categories of their content [45]. Cybersecurity experts have also been investigating the use of kNN to detect cyberattacks in real time. This method has been used to sense attacks like a fake data injection attack and it also works fine when the data can be constituted by a model that measures the distance to other data.

Support Vector Machines

A support vector machine (SVM) comes under the linear regression model, it classifies the data points specifically. In other words, it is an addition of linear regression which discovers a plan which separates data into two different classes. SVMs are used for the classification of cyberattacks. These are heavily used in web applications. Whenever a user opens their web browser the SVM automatically starts and keeps running in the background and monitors every page the user has been to. In the background, the SVM matches for vulnerabilities by observing various properties of the threat. After this, it detects the type of threats and classifies the cyberattack [46]. Moreover, this technique is also used to analyse the traffic patterns on the internet and separate them into their classes, like FTP, HTTP, etc. These SVMs are also widely used in applications where the attacks can be simulated. Moreover, as a SVM is a very good tool for classifying cyberattacks it has its drawbacks too, like it is very sensitive to attacks, it can only classify into two classes at a time and its performance is based on kernel functions.

Artificial Neural Networks

An artificial neural network (ANN) is a part of AI which is meant to imitate a human brain. It has a many artificial neurons interconnected via nodes much like a human brain, in which many neurons are interconnected to each other. These artificial neurons are known as processing units. These processing units are divided into two parts: input unit (input layer) and the output unit (output layer). It actually has three layers inside it, but the third layer is hidden and so it is called a hidden layer as well. This hidden layer is between the input layer and output layer [47]. ANNs can be trained any way, either by supervised learning, unsupervised learning or reinforcement learning. As most of the area under an intrusion detection system (IDS) relies on AI techniques this makes ANN the solid IDS and ANNs have also performed well towards DoS attacks [47]. There are three types of IDS based on ANN, they are: host-based IDS, network-based IDS and vulnerability-assessment IDS. Several studies have also suggested that use of ANN can establish pattern recognition and recognise an attack in circumstances where the rules are unknown. Apart from this, the only disadvantage here is that the ability of ANN to identify signs of intrusion depends on the training requirements of the data and methods, the use of which is very critical. Mohammad et al. [48] discussed about the effectiveness of FDIA on DC microgrids networks consisting of a parallel DC–DC converter and controlled by leakage-based control strategies using ANN. In other studies, Jonghoon et al. [49] developed an AI-SIEM system based on ANN methods like convolutional neural networks (CNN), fully connected neural networks (FCNN) and long short term memory (LSTM); datasets CICDS20017 and NSLKDD are used for evaluation. Khan et al. [50] devised the use of ANN for malware classification and have used the dataset provided by the Microsoft Malware Classification Challenge. In another work, Xavier et al. [51] suggested the performance evaluation of various proven ML algorithms commonly used in IDS. Similarly, Riaz et al. [52] proposed a network IDS built on convolutional ANN algorithms and have used the KDD99 dataset for evaluation.

QUANTUM-RESISTANT CYBERSECURITY TECHNOLOGIES

As the field of quantum computers is progressing so rapidly, it is also increasing security threats. As of today, the quantum computers that exist are not capable of breaking encryption methods but in the near future new types of computers based on quantum physics can break most modern encryption systems. So, due to these threats, quantum-resistant cybersecurity technologies are being created.

MULTIVARIATE CRYPTOGRAPHY

Asymmetric cryptographic algorithms hinged on multivariate polynomials over a finite field F are known as multivariate cryptography. The main assumption of safety is supported by the NP-hardness of the problem for solving nonlinear equations in a finite field. The same basic structure is used for creating all the multivariate cryptosystem. The multivariate quadratics consist of two keys: public and private. The unbalanced oil and vinegar (UOV) cryptosystem is a type of multivariate cryptography system. It is used for signatures. Apart from UOV there are other examples of this cryptosystem: multivariate public key cryptography (MPKC), TTS, rainbow, QUARTZ, etc.

CODE-BASED SIGNATURES

Code-based cryptography includes all symmetrical or asymmetrical cryptographic systems, the security of which depends on the difficulty of the decoding in the linear error correction code, probably selected with a certain structure or family. These systems are based on the principle of error-correcting codes for the construction of one-way function, and these are also contemplated to be the next generation of cryptographic algorithms for standard public key systems in current use. Adding a few errors in the input and creating some error patterns through message encoding are a few of the measures taken into account while encrypting in this cryptosystem. For decryption, the errors are removed from the input and the plain text is extracted from the errors.

Hash-Based Signatures

Cryptographic algorithms based on the security of hash functions are known as hash-based cryptosystems. These hash-based signatures use one-time signatures scheme. These one-time signatures can be used for particularly every single message. These hash-based signatures are highly flexible as they can be implemented in addition to any hash function that has basic security requirements. They also have parameters which define aspects like key size and sighing speed which allows trade-offs in limited environments. The ease of implementation and customisation of hash-based signatures makes them a good representative for the IoT ecosystems. Some of the examples of hash-based signature schemes are Merkle's scheme, XMSS, SPHINCS, etc.

Lattice-Based Cryptography

Cryptographic algorithms which use lattices in construction or in the security proof are called lattice-based cryptography. Similar to hash-based schemes, these have also proved to be a good approach in post-quantum cryptography. These schemes provide safety tests based on NP-hardness problems of medium to the worst hardness. Then in order to be secure in the quantum age, the lattice-based cryptography also excels in terms of performance, largely due to their inherited linear algebra-based vector actions on integers. Finally, the lattice-based cryptography schemes offer enhanced functions for greater security services such as identity-based cryptography, classic cryptography and many more.

CONCLUSION

As an evolving technology the IoT has linked numerous devices to the internet, vividly revolutionising our daily lives. Cars, trucks, consumer goods, industrial parts and equipment and other more everyday objects combined with great analytical capabilities and the internet, promise to change the world we live in today. With this much popularity, several risks also arise, and due to the lower level of security and the growing popularity of the IoT, a diversity of attacks is being carried out and more new attacks are also being discovered against the IoT. Cyber experts assure that with proper security measures in place, the IoT will become an outstanding network. In this chapter, we have discussed the layer of cybersecurity-oriented IoT architecture, some popular types of attacks or techniques like DoS, physical attacks and so on, which are used by attackers against the IoT together with their countermeasures and prevention methods, the role of AI-ML and its branches in cybersecurity like their uses in threat detection and analysis. Furthermore, we have discussed some cybersecurity technologies used for quantum resistance, like hash-based cryptosystems, code-based cryptography, etc.

REFERENCES

1. Khaleeque, R. & Mansoor, H. (2020). Internet of things (iot) and it's needs. *Global Sci-Tech*, *12*(1), 38. doi: 10.5958/2455-7110.2020.00006.3

2. Behmann, Fawzi & Wu, Kwok (2015). Impact of C-IoT and tips. In Fawzi Behmann & Kwok We (eds.) *Collaborative Internet of things (C-IoT): For future smart connected life and business* (pp. 239–251). Wiley, Hoboken.
3. Saxena, S., Bhushan, B. & Ahad, M. A. (2021). Blockchain based solutions to Secure Iot: Background, integration trends and a way forward. *Journal of Network and Computer Applications*. doi: 10.1016/j.jnca.2021.103050, PubMed: 103050
4. Bhushan, B. & Sahoo, G. (2020). *Requirements, protocols, and security challenges in wireless sensor networks: An industrial perspective. Handbook of computer networks and cyber security* (pp. 683–713). Springer, Cham. doi: 10.1007/978-3-030-22277-2_27
5. Goyal, S., Sharma, N., Bhushan, B., Shankar, A. & Sagayam, M. (2020). IoT enabled technology in secured healthcare: Applications, challenges and future directions. *Cognitive Internet of Medical Things for Smart Healthcare Studies in Systems, Decision and Control.*, 25–48. doi: 10.1007/978-3-030-55833-8_2
6. Sethi, R., Bhushan, B., Sharma, N., Kumar, R. & Kaushik, I. (2020). Applicability of industrial IoT in diversified sectors: Evolution, applications and challenges. *Studies in Big Data Multimedia Technologies in the Internet of Things Environment*, 45–67. doi: 10.1007/978-981-15-7965-3_4
7. Lu, Y. & Xu, L. D. (2019). Internet of things (IoT) cybersecurity research: A review of current research topics. *IEEE Internet of Things Journal*, 6(2), 2103–2115. doi: 10.1109/JIOT.2018.2869847
8. Bhushan, B. & Sahoo, G. (2017). Recent advances in attacks, technical challenges, vulnerabilities and their countermeasures in wireless sensor networks. *Wireless Personal Communications*, 98(2), 2037–2077. doi: 10.1007/s11277-017-4962-0
9. Arora, A., Kaur, A., Bhushan, B. & Saini, H. (2019). Security concerns and future trends of Internet of things In *2nd International Conference on Intelligent Computing, Instrumentation and Control Technologies (ICICICT)*.
10. Kambourakis, G., Kolias, C. & Stavrou, A. (2017). The Mirai botnet and the IoT zombie armies. In *IEEE Military Communications Conference (MILCOM)*.
11. Han, W. & Xiao, Y. (2017). Cybersecurity in internet of things (iot). *Big Data Analytics in Cybersecurity*, 221–244. doi: 10.1201/9781315154374-10
12. Madaan, G., Bhushan, B. & Kumar, R. (2020). Blockchain-based cyberthreat mitigation systems for smart vehicles and industrial automation. *Studies in Big Data Multimedia Technologies in the Internet of Things Environment*, 13–32. doi: 10.1007/978-981-15-7965-3_2
13. Sinha, P., Jha, V. K., Rai, A. K. & Bhushan, B. (2017). Security vulnerabilities, attacks and countermeasures in wireless sensor networks at various layers of OSI reference model: A survey In *International Conference on Signal Processing and Communication (ICSPC)*.
14. Wang, C., Wang, D., Tu, Y., Xu, G. & Wang, H. (2020). Understanding node capture attacks in user authentication schemes for wireless sensor networks. *IEEE Transactions on Dependable and Secure Computing*, 1. doi: 10.1109/tdsc.2020.2974220
15. Li, L., Chen, Y., Wang, D. & Zheng, T. F. (2017). A study on replay attack AND Anti-Spoofing for Automatic Speaker verification. *Interspeech*. doi: 10.21437/interspeech.2017-456
16. Belous, A. & Saladukha, V. (2020). Computer viruses, malicious logic, and spyware. *Viruses, Hardware and Software Trojans*, 101–207. doi: 10.1007/978-3-030-47218-4_2
17. Hund, R., Willems, C. & Holz, T. (2013). Practical timing side channel attacks against kernel Space ASLR. In *IEEE Symposium on Security and Privacy*.
18. Basim, H. & Ahmed, T. (2018). An improved strategy for detection and prevention ip spoofing attack. *International Journal of Computer and Applications*, 182(9), 28–31. doi: 10.5120/ijca2018917667

19. Li, Q., Li, H., Wen, Z. & Yuan, P. (2017). Research on the p2p Sybil attack and the detection mechanism In *8th IEEE International Conference on Software Engineering and Service Science (ICSESS)*. doi: 10.1109/icsess.2017.8343002

20. Bhushan, B. & Sahoo, G. (2017). Detection and defense mechanisms against wormhole attacks in wireless sensor networks In *3rd International Conference on Advances in Computing, Communication & Automation (ICACCA)*.

21. Babbar, G. & Bhushan, B. (2020). Framework and methodological solutions for cyber security in Industry 4.0. *SSRN Electronic Journal*. doi: 10.2139/ssrn.3601513

22. Vayansky, I. & Kumar, S. (2018). Phishing – Challenges and solutions. *Computer Fraud and Security, 1*(1), 15–20. doi: 10.1016/s1361-3723(18)30007-1

23. Xue, L. & Hu, Z. (2015). Research of worm intrusion detection algorithm based on statistical classification technology. In *8th International Symposium on Computational Intelligence and Design (ISCID)*.

24. Khan, H. A., Syed, A., Mohammad, A. & Halgamuge, M. N. (2017). Computer virus and protection methods using lab analysis In *2nd International Conference on Big Data Analysis (ICBDA)*. IEEE Publications

25. Mallikarajunan, K. N., Preethi, S., Selvalakshmi, S. & Nithish, N. (2019). Detection of spyware in software using virtual environment In *3rd International Conference on Trends in Electronics and Informatics (ICOEI)*.

26. Oh, S., Bae, H., Yoon, S., Kim, H. & Cha, Y. (2016). Malicious script blocking detection technology using a local proxy. In *10th International Conference on Innovative Mobile and Internet Services in Ubiquitous Computing (IMIS)*.

27. Gudipati, V. K., Vetwal, A., Kumar, V., Adeniyi, A. & Abuzneid, A. (2015). Detection of Trojan horses by the analysis of system behavior and data packets. In *2015 Long Island Systems, Applications and Technology*. pp. 1–4, doi: 10.1109/LISAT.2015.7160176.

28. Meneghello, F., Calore, M., Zucchetto, D., Polese, M. & Zanella, A. (2019). Iot: Internet of threats? A survey of practical security vulnerabilities in real iot devices. *IEEE Internet of Things Journal, 6*(5), 8182–8201. doi: 10.1109/jiot.2019.2935189

29. Tsiknas, K., Taketzis, D., Demertzis, K. & Skianis, C. (2021). Cyber threats to industrial iot: A survey on attacks and countermeasures. IoT, *2*(1), 163–186. doi: 10.3390/iot2010009

30. Bhushan, B., Sahoo, G. & Rai, A. K. (2017). Man-in-the-middle attack in wireless and computer networking — A review In *3rd International Conference on Advances in Computing, Communication & Automation (ICACCA)*.

31. Rathi, R., Sharma, N., Manchanda, C., Bhushan, B. & Grover, M. (2020). Security challenges & controls in cyber physical system. In *9th International Conference on Communication Systems and Network Technologies (CSNT)*. IEEE Publications.

32. The iot ransomware threat is more serious than you think (n.d.). Retrieved from https://www.iotsecurityfoundation.org/the-iot-ransomware-threat-is-more-serious-than-you-think/

33. Kaspersky (2021, January 13). Brute force attack: Definition and examples. Retrieved from https://www.kaspersky.com/resource-center/definitions/brute-force-attack

34. Kammara, T. & Moh, M. (2019). Identifying iot-based botnets. *Botnets*, 293–326. doi: 10.1201/9780429329913-9

35. Bhayo, J., Hameed, S. & Shah, S. A. (2020). An efficient counter-based ddos attack detection framework leveraging software defined iot (sd-iot). *IEEE Access*, 8-221631. doi: 10.1109/access.2020.3043082

36. Lee, I. (2020). Internet of things (iot) cybersecurity: Literature review and iot cyber risk management. *Future Internet, 12*(9), 157. doi: 10.3390/fi12090157

37. Austin, G. (2018). Corporate cybersecurity. *Cybersecurity in China*, 65–79. doi: 10.1007/978-3-319-68436-9_4

38. Kuzlu, M., Fair, C. & Guler, O. (2021). Role of artificial intelligence in the internet of things (iot) cybersecurity. *Discover Internet of things*, *1*(1). doi: 10.1007/s43926-020-00001-4

39. Clark, G., Doran, M. & Glisson, W. (2018). A malicious attack on the machine learning policy of a robotic system In *17th IEEE International Conference on Trust, Security and Privacy in Computing and Communications/12th IEEE International Conference on Big Data Science and Engineering (TrustCom/BigDataSE)*, pp. 516–521.

40. Xin, Y., Kong, L., Liu, Z., Chen, Y., Li, Y., Zhu, H., … Wang, C. (2018). Machine Learning and deep learning methods for cybersecurity. *IEEE Access*, *6*, 35365–35381. doi: 10.1109/ACCESS.2018.2836950.

41. Azwar, H., Murtaz, M., Siddique, M. & Rehman, S. (2018). Intrusion Detection in secure network for Cybersecurity systems using Machine Learning and Data Mining. In *5th International Conference on Engineering Technologies and Applied Sciences (ICETAS)*, pp. 1–9. IEEE Publications.

42. Zhang, S., Xie, X. & Xu, Y. (2020). A Brute-Force Black-Box method to attack machine learning-based systems in cybersecurity. *IEEE Access*, *8*, 128250–128263. doi: 10.1109/ACCESS.2020.3008433.

43. Li, Y., Xiong, K., Chin, T. & Hu, C. (2019). A machine learning framework for domain generation algorithm-based malware detection. *IEEE Access*, *7*, 32765–32782. doi: 10.1109/ACCESS.2019.2891588.

44. Ortiz Garcés, I., Cazares, M. F. & Andrade, R. O. (2019). Detection of phishing attacks with machine learning techniques in cognitive security architecture. In *2019 International Conference on Computational Science and Computational Intelligence (CSCI)*, pp. 366–370. doi: 10.1109/CSCI49370.2019.00071.

45. Jha, S., Kumar, R., Hoang Son, L., Abdel-Basset, M., Priyadarshini, I., Sharma, R. & Viet Long, H. (2019). Deep learning approach for software maintainability metrics prediction. In *IEEE Access*, *7*, 61840–61855.

46. Sharma, R., Kumar, R., Sharma, D. K., Le Hoang, S., Priyadarshini, I. & Pham, B. T. (2019). Dieu Tien Bui & Sakshi Rai. Inferring air pollution from air quality index by different geographical areas: Case study in India. *Air Quality, Atmosphere and Health*, *12*(11), 1347–1357.

47. Sharma, R., Kumar, R., Singh, P. K., Raboaca, M. S. & Felseghi, R.-A. (2020). A systematic study on the analysis of the emission of CO, CO_2 and HC for four-wheelers and its impact on the sustainable ecosystem. *Sustainability*, *12*(17), 6707.

48. Habibi, M. R., Baghaee, H. R., Dragičević, T. & Blaabjerg, F. (2021). False Data Injection Cyber-Attacks Mitigation in Parallel DC/DC Converters Based on Artificial Neural Networks. *IEEE Transactions on Circuits and Systems II: Express Briefs*, *68*(2), 717–721. doi: 10.1109/TCSII.2020.3011324.

49. Lee, J., Kim, J., Kim, I. & Han, K. (2019). Cyber threat detection based on artificial neural networks using event profiles. *IEEE Access*, *7*, 165607–165626. 10.1109/ACCESS.2019.2953095.

50. Khan, M., Baig, D., Khan, U. S. & Karim, A. (2020). Malware classification framework using convolutional neural network. In *2020 International Conference on Cyber Warfare and Security (ICCWS)*, pp. 1–7. doi: 10.1109/ICCWS48432.2020.9292384.

51. Larriva-Novo, Xavier A., Vega-Barbas, M., Villagrá, V. A. & Sanz Rodrigo, M. (2020). Evaluation of cybersecurity data set characteristics for their applicability to neural networks algorithms detecting cybersecurity anomalies. *IEEE Access*, *8*, 9005–9014. doi: 10.1109/ACCESS.2019.2963407.

52. Khan, Riaz U., Zhang, X., Alazab, M. & Kumar, R. (2019). An improved convolutional neural network model for intrusion detection in networks. In *2019 Cybersecurity and Cyberforensics Conference (CCC)*, pp. 74–77, doi: 10.1109/CCC.2019.000-6.

5 Blockchain-Based Cybersecurity Solutions for Industry 4.0 Applications

Tanya Shrivastava and Bharat Bhushan

CONTENTS

DOI: 10.1201/9781003097518-5

INTRODUCTION

Industry 4.0 is revolutionary. It has introduced some concepts like cloud computing and the Internet of Things (IoT) that have had a major impact in changing the global market [1]. Industry 4.0 describes data exchange manufacturing and its use in automation. According to the Boston Consulting Group, nine technologies shape Industry 4.0 and those are: simulation, data and analytics, cybersecurity, the IoT, cloud, augmented reality, robots, system integration and additive from manufacturing [2]. These technologies together create a 'smart factory' in which there is communication between humans, machines and systems that helps in the monitoring and coordination of processes [3]. Industry 4.0 uses the IoT for the process of digital manufacturing. In recent years huge efforts have been applied in shaping Industry 4.0. The 'Fundamental Principles of Industry 4.0' were introduced which basically consists of some parameters such as extensive internet use, process virtualisation, flexibility of production, etc. [4]. These emphasise Industry 4.0 to discover new technologies instead of focusing on traditional devices or gadgets. Extensive internet use means increasing capacity by developing new technologies while keeping the internet as an information source [4]. Industry 4.0 is trying to bring the physical and virtual worlds together in a networked global system [5].

Cybersecurity should be one of the major concerns in this expanding technology, to prevent any disclosure of sensitive authentic information [6]. Traditional security parameters are not enough to preserve the safety of the system, thus development of new technology and its implementation is necessary to minimise the threat of attacks [7]. Therefore, this is the purpose of the cybersecurity infrastructure of the IoT, to protect billions of network devices from malicious intentions [8, 9]. As the network of devices is vast, consisting of heterogenous devices such as mobiles, televisions, computers, sensors, etc., these also affect Industry 4.0 as day by day new network technologies are taking part in IoT-based cybersecurity [10].

Industry 4.0 refers to the conversion of traditional technologies or factories into smart ones, so they become more efficient in handling resources. Industry 4.0 mainly focuses on the collection or gathering of data from diverse sources [11]. It should take place in such an environment that permits it to run various processes on it by the factory devices. Industry 4.0 also proposes the use of autonomous communication between the devices and the IoT and proposes the development of smart cities [12]. Smart cities have made daily living more comfortable by helping in traffic reduction and environmental benefits. Sensors, vehicles, etc. are some components of smart cities and the IoT. But this also makes the smart city vulnerable to cyberattacks [13]. Attackers can have malicious intentions towards smart vehicles; they can cut the power or water by hijacking the system. They can cause accidents and endanger life by compromising the traffic lights [14]. In summary, the major contributions of this work can be enumerated as follows:

- A description of the IoT network and the major security dangers or attacks it could encounter following the layered security architecture for the IoT.
- A detailed explanation of all the major cybersecurity threats that Industry 4.0 can suffer and later presents the related cybersecurity solutions.
- A description about the need for blockchain technology and smart contracts for Industry 4.0 applications.
- How Industry 4.0 technologies benefit from the use of blockchain technology and the related application areas.

The remainder of the chapter is organised as follows. The next section presents the four-layered cybersecurity architecture for the IoT and describes how the network, application, middleware and session layer operates. This is followed by a description of all the major cybersecurity threats to which Industry 4.0 is vulnerable and the ways it can be affected by them. All the cybersecurity solutions for Industry 4.0. factories are then described. The following section presents the need for the use of blockchain technologies for the benefit of Industry 4.0. This concludes with a presentation of the way in which Industry 4.0 benefits from blockchain technology. Thus, high security parameters should be set for the protection of smart cities and the people living in them.

FOUR-LAYERED CYBERSECURITY-ORIENTED IoT ARCHITECTURE

The IoT is global, and in these networks smart devices also participate like sensors or smartphones, thus making them vulnerable to cyberattacks such as denial-of-service (DoS) because of some limitations like storage capacity or power consumption, etc. A DoS is an attack which disrupts the normal functioning of an IoT device and prevents other users from accessing it [15]. In the four-layered cybersecurity-oriented architecture for the IoT, the framework is divided into four layers: sensing layer, networking layer, middleware layer and application layer. Confidentiality, integrity, availability and integration are some essential parameters to make the architecture more efficient [16]. The architecture design should maintain memory, embed software, consume energy, etc. in equilibrium. Cybersecurity policies may change thus they need an architecture that assists devices for dynamic interaction. Every layer in this architecture has some vulnerability that the attackers can exploit. The architect is concerned with many things, such as some protocols, principles and networking and communications, etc. [17–19]. Four-layered cybersecurity architecture is depicted in Figure 5.1.

```
┌─────────────────────────┐
│                         │   ◄──── Sensing data, collect and transmit data to network
│      Sensing Layer      │
│                         │
└─────────────────────────┘
            ▲
┌─────────────────────────┐
│                         │   ◄──── Carries and transmit information to IoT devices
│      Network Layer      │
│                         │
└─────────────────────────┘
            ▲
┌─────────────────────────┐
│                         │   ◄──── Collects and filters the data
│     Middleware Layer    │
│                         │
└─────────────────────────┘
            ▲
┌─────────────────────────┐
│                         │   ◄──── Provides user end functionality
│    Application Layer     │
│                         │
└─────────────────────────┘
```

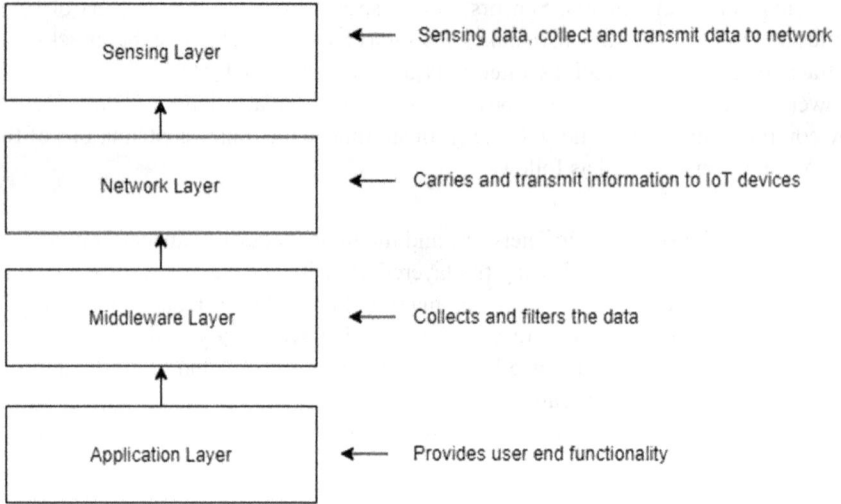

FIGURE 5.1 Four-layered cybersecurity-oriented architecture for the IoT.

SENSING LAYER

It works similar to human ears, nose and eyes. It identifies things and then collects information, processes and sends it to the entire network with the help of attached data sensors. Some threats that the sensing layer can receive are eavesdropping, exposure of sensor nodes, fake nodes, timing or replay attacks, node capture and data attack. A replay attack is a cyberattack in which an intruder eavesdrops on the conversation between two parties i.e., sender and receiver, and takes information from the sender and then delays or resends it [20]. It is also called a play-back attack. Attackers do not need any special skill for this attack to decrypt the information. To prevent this attack, one should have the right type of encryption. In the timing attack, the attacker discovers vulnerabilities and extracts secrets maintained in the security of a system by observing how long it takes the system to respond to different queries, input or cryptographic algorithms. A node capture attack is when the intruder takes over a node and extracts authentic information and data from it [21]. Other than side channel, time consumption, power consumption and electromagnetic radiation are also a threat for this layer.

NETWORK LAYER

It is also known as a transmission layer and acts as a medium between the application layer and the sensing layer. The main function of this layer is to carry and transmit the information collected from physical objects through sensors to different IoT devices over the network [22]. The medium of transmission can be guided or unguided media between heterogenous smart devices. Cloud computing, internet gateways, switching, etc. are performed at this layer with the help of some of the latest technologies.

It connects smart devices and networks to each other and therefore it is sensitive to attacks. The main cybersecurity issues at this layer are confidentiality, privacy and compatibility. The network layer is mostly vulnerable to man-in-the-middle attacks. A man-in-the-middle attack is when the attacker positions themself in the conversation between the sender and the receiver to eavesdrop or impersonate one of them. Other things that the network layer is most likely vulnerable to are spoofing, modification, replay attacks and sybil attacks. A sybil attack is an attack in which a node in the network operates multiple identities actively at the same time and undermines the authority/power in reputation systems [24, 25]. Facebook and twitter are vulnerable to sybil attacks [23].

Middleware Layer

The middleware layer is in between the network layer and the application layer. It eases the integration of new legacies and hides the complexities of the lower layers [30]. If the data has been exchanged then the middleware layer operates and manages the integrity, authenticity and confidentiality of it. It is based on service-oriented architecture (SOA). It collects and filters data received from the hardware devices then it performs information recovery on that and handles access control to the application devices [31]. Some of the attacks that the middleware layer is vulnerable to are malicious insider, underlying infrastructure, third-party relationship and virtualisation. A turncloak attack is also a name given to malicious insider attacks. In this an internal attacker modifies and extracts data within the network on purpose. An underlying attack is a platform-as-a-service based attack. In this the aim of the developer is to keep an IoT application secure. Third-party attacks are caused by third-party entities such as mashups that increase the data security. Hyperjacking is a kind of virtualisation attack. A virtualisation attack is when a virtual machine can cause damage or affect other virtual machines. An insider may also get unauthorised access to the system or to a network and may find the vulnerable places through investigation [32].

Application Layer

This layer defines the applications which have been applied to the IoT and it goes through the entire system functionality for the user and provides service to the application. The application can be anything such as a smart home, smart cities, etc. The information is collected via sensors and therefore the service can vary. Some common security problems that the application layer faces are data access, protection and recovery, malicious code attacks, cross site scripting, the ability of dealing with mass data and some software vulnerabilities. Cross-site scripting is a type of injection attack in which malicious code script is inserted into a trusted website, while a malicious code attack consists of a harmful code that causes damage to the system and activate itself. The attacker inserts malware in the form of a virus, worm, etc. The ability of mass data explains why several devices and the connection between them lack the ability to deal with processing data as per the requirement and thus

TABLE 5.1

Summary of Cybersecurity-Oriented IoT Architecture

Layer	Functionality	Attacks
Sensing layer	• Collects and transmits information to IoT devices. • Connects devices and networks to each other.	Eavesdropping, replay/timing attack, fake nodes.
Network layer	• Eases integration and hides complexities. • Manages the integrity, authenticity and confidentiality of data.	Man-in-the-middle, spoofing, sybil attack, replay attack.
Middleware layer	• Provides user end functionality. • Provides service to the application.	Third party attack, malicious insider, underlying attack, hyper jacking.
Application layer	• Provides user end functionality. • Provides service to the application.	Cross site scripting, malicious code attack.

cause network disturbance [33, 34]. A brief description of the aforementioned layers is presented in Table 5.1.

SECURITY THREATS IN INDUSTRY 4.0

Cyberthreats are major risks for the development of Industry 4.0 affecting it both financially and technically. In the previous few years, Industry 4.0 has suffered billions of dollars of damage because of cyberattacks which affected the production rate of factories. As Industry 4.0 is rapidly growing, it is continuously suffering from security and privacy threats. Some major security threats to Industry 4.0 are discussed in the subsections below.

CYBERESPIONAGE

It is the act of gaining unauthorised information from the information-holder using various internet methods. It is widely used to obtain access to sensitive government or corporate information. Various well-organised cybercriminal groups have targeted Industry 4.0 with cyberattacks to steal sensitive information. One famous group is the Black Vine group, and it is well-known for targeting the aerosol and healthcare industries, etc. Two other cyberespionage groups like Aurora and GhostNet are famous for recent attacks [26, 27]. Corporate and product data are very sensitive and thus their theft is very common [28, 29].

DENIAL-OF-SERVICE

Denial-of-service (DoS) or distributed denial-of-service (DDoS) describe cyberthreats that cause a machine or network to shut down or become unavailable for the intended users [26]. Most of the systems are interconnected in an environment, so the non-availability of some gadgets makes it critical for the production environment and therefore results in making these attacks popular [27]. Flooding or crashing services are two main methods to accomplish DoS attacks. Some popular flood attacks are SYN, ICMP and buffer overflow attacks [28], while other DoS attacks just exploit the vulnerabilities in a system to crash it. These attacks are carried out by devices that are affected by malware and allow it to be controlled by the intruder. These devices are known as bots or group of bots and known as a botnet. When the botnet targets the victim's system, each bot transmits a request to the victim's IP address and results in the flooding of the network [29]. The usual symptoms of a DoS attack are the slowing down of the network, difficulty in accessing files, website unavailability and inability to access it [30].

Supply Chain and Extended Systems

The Industry 4.0 paradigm features various connections among several organisational environments in order to help the supply chain become more efficient. However, the supply chain consists of some vulnerabilities that attackers can find and take advantage of. A single vulnerability can cause many attacks such as phishing or the theft of credentials, and that can result in mass data exposure in a factory.

Smart Security and Smart Factory

The majority of factories are unaware of the risk that happens with the paradigm of Industry 4.0, their responsibility is to handle the security issues that arise when a serious problem happens. But technical products alone cannot handle these problems. The human factor is necessary. Awareness of security among employees is an important aspect, involving operators to software to engineers. Awareness can take various forms, such as campaigns that raise awareness and involve entire research groups in the manufacturing sector or in higher education institutes, and who discuss cybersecurity and deliver guidelines to experts.

Advanced Persistent Threat

Advanced persistent threat (APT) is a term used to describe the specific type of attacks in which the attacker or a group of attackers gains an illegal and long-term presence inside a network to steal highly sensitive data, with potentially critical consequences [31]. Huge effort is needed to carry out these attacks, so the targets chosen are of high value, like government networks or large corporations. These attacks can be broken down into three phases, namely infiltration, expansion and extraction [32]. Some of the examples of APT are DragonFly, ExPert, Cozy Bear, Ocean Buffalo, etc.

Logistics

This field consists of many wireless devices like our mobile phones. The cyber-threats here can be caused due to installation of malicious programmes, connection to a non-encrypted Wi-Fi or even to an unknown Wi-Fi source and by using business protocols which use the HTTP instead of HTTPS protocol. These vulnerabilities can reveal a lot of information like the serial number of the device, information about the database, IT infrastructure and more which is stored on the device or whatever the device is operating at that moment to the attacker. One well-known example of a virus affecting Android devices is the HummingBad virus. It attracts users from click-fraud advertisements. It collects personal data form our devices and sells the data on.

ACTIVE SUPERVISION

This area comes under corporate networks and it mainly includes desktop hardware. The defects that can cause cyberthreats that are very similar to the common faults we regularly face, like computers not being locked after the work is done, not having an updated antivirus software or even an antivirus at all, not having updated digital signatures, installation of pirated or unlicenced software or accessing untrusted and unauthorised websites, direct connection to external devices, etc. All these vulnerabilities can be caused due to threats like a phishing attack, malware, worms, the transfer of data onto unauthorised external devices, data theft or loss, etc.

CYBERSECURITY SOLUTIONS FOR FACTORIES

In this section we describe some solutions to shield against an attack or security threat that affects Industry 4.0 factories. These are some existing solutions that are used by factories for protection from malicious intentions. We describe some short- and long-term solutions to defend against threats. Short-term approaches are called countermeasures while long-term ones are known as solutions. These countermeasures should be updated frequently for better protection and efficiency.

CONTROLLED INDUSTRIAL ENVIRONMENT RESEARCH

Both logical and physical solutions are required by modern factories for the purpose of information security. These are virtual and they have an interface, so they are able to come out from the system and compromise it. The interface can be living, and the breach appears at the time when the information is being exchanged. The chances of information leakage have increased since the massive growth of the wireless network. For this purpose, physical information security has been created. The solutions are in the form of 'protected areas'. The solution comes with various objectives such as to stop giving access to unauthorised people, preventing authentic information spreading at protected area walls, controlled and protected system. Three levels of solutions are there in order to achieve these goals. The first layer consists of conventional methods that are used for protected facilities and the screening of people. This layer helps in limiting access so that only authorised people can have it, but there remain some risks. The second layer helps in preventing the information spreading to outside from the protected area zone with the help of physical characteristics such as visual or audible alerts using visual and sound shielding. The third layer helps in maintaining the security proposal context.

SOFTWARE DEFINED NETWORKING (SDN)

SDN helps in the improvement of data manufacturing transparency for corporate applications. Sensor data can compute a performance indicator and monitor the machine status and manufacturing process quality. IT system interconnectivity enhancement also exposes devices to cyberattacks that can infect other systems. Some common attacks that can happen with Manufacturing Execution System (MES) are related to network scanning or probing in which defence-in-depth is an effective countermeasure. A countermeasure is a term that represents actions to prevent threats and can also report a threat. Usually, multiple firewall segments are inserted in between network architecture to minimise the threats as proposed by cybersecurity standards. Configuration rules should be defined in secure manner and SDN helps in this regard. It provides an overview of the security architecture that helps administrators and also permits users in programmatically controlling the network architecture. It consists of three objectives as segment creations without the reconfiguration of existing networks, unidirectional access mechanisms development and loophole reduction.

DIRECT-TO-MACHINE APPROACH IN CPS

One of the major concerns in the security of Industry 4.0 information technology (IT) is long value chains. During the production time, various kinds of data are collected, and they are used for quality checks and maintenance, even though only part of it is essential for security. The area of focus for IoT solutions should be the materials bill, information of design and control parameters. The architecture that is currently in use is separated from the production environment and when the data leave the server there is zero control over subcontractors. Once the data are received the operators are left unsupervised and then many problems arise such as damage to the data. A new way is proposed that it limits and helps in information protection flow in devices for security.

ENSEMBLE INTELLIGENCE IN ADVANCED MANUFACTURING

Traditional cybersecurity architecture mainly focuses on ideas that help in providing confidentiality, integrity, access control, authentication, etc. to prevent cyberattacks and threats. There is also an algorithm for the detection of cyberattacks to help Industry 4.0 systems and some other systems that are internet-driven. Generally, algorithms are needed by cyberattack detection systems that aim to work on the collected data by many events that occur in a cyberenvironment. Furthermore, poor accuracy can have an impact on the performance of the system in a negative manner and can also result in security complications such as unnoticed intrusions or false alarms. To effectively differentiate between normal and abnormal data, technologies such as deep learning or machine learning are used by systems. Another proposed algorithm is the neural network oracle (NNO) classification algorithm, and it consists of three components, these are the neural network, neural network oracle and genetic algorithms.

BEHAVIORAL MODELS AND CRITICAL STATE DISTANCE

Since the beginning of the 21st century, the industrial control system (ICS) has been a constant target for cyberattacks by hackers because of the damage they can cause on the system and the environment. The ICS network can be categorised into three layers: level 0 that is the operative part containing sensors and actuators, level 1 that is the control part consisting of PLC or HMI then there is level 3 that is for supervision with control rooms or SCADA [33]. Some networking solutions like firewalls or DMZ are used in levels 2 and 3 in need of protection from attacks such as DDoS and man-in-the-middle attacks. The solution in this layer is helpful because of its similarities with traditional IT infrastructure. But for layers 0 and 1, these solutions do not work because they have their own attacks like random attacks or false data injection. Some other attacks also are temporal, sequential, direct and over-soliciting. Another methodology solution is dedicated to the protection of low-level elements. This way is considered as the final shield for the protection of the system from cyberattacks [34].

BLOCKCHAIN FOR INDUSTRY 4.0 APPLICATIONS

This section describes how Industry 4.0 applications benefit from blockchain technology. Industry 4.0 needs the help of blockchain to deal with various issues it faces with its development, such as how to implement networks, the cooperation of clients, suppliers and manufacturers, smart industry and the deployment of multiple technologies [35, 36].

NEED FOR BLOCKCHAIN

Blockchain brings many benefits for industrial development and it is also useful for Industry 4.0 applications. But we cannot use blockchain every time for every problem. As an example, in private networks some powerful tools are provided by some traditional databases. Though there are also some other technologies which are similar to blockchain technology, they can be helpful for Industry 4.0 applications, as such for IoT applications, etc. We can take an example of tangle which is a directed acyclic graph (DAG) implementation and can be used for modelling various kinds of information. It does not make use of blocks and requires no miners and therefore provides transactions at high speed with zero network fees [37, 38].

Tangle also possess some downsides such as the technology it uses is not proven to be blockchain because of the lack of users, it cannot provide decentralized applications (DApp) functionality and Turing-complete solutions, it may have some security issues as it is vulnerable to some attacks such as a 34-percent attack and its decentralisation is hindered because of the use of coordination nodes for issues such as synchronisation by IoTA. thus while deciding which technology should be used in between blockchain and the alternative options, Industry 4.0 applications should recognise certain features. In the industrial field, IoTA is the most popular DAG-based project. A brief amount of proof-of-work (PoW) is necessary for the verification of the two other transactions in an IoTA system [39]. Whenever decentralisation is needed in applications, blockchain can be helpful in playing a big role. Decentralisation can be used by Industry 4.0 applications when the centralised system is not trusted. Blockchain is not needed when there is trust among different entities.

When there are payment needs to perform then trust becomes necessary. Payments can also be processed through some traditional payment systems, but they have two drawbacks as they have high transaction fees, and they need to be trusted blindly [40]. The issued raised in public transactions are mainly trust and transparency. These also contain information that can be exposed to the public. This is followed in a strict manner in some Industry 4.0 applications for accurate records and to make use of data techniques, etc. [41]. Databases have traditionally already provided these features and their security is necessary. Since they are accessed through the internet it makes them more vulnerable to attacks such as attacks on data privacy, etc. Industry 4.0 applications may require other features such as P2P communications for data exchange for industrial processes [42].

Desai et al. [43] proposed a hybrid blockchain model consisting of both public and private blockchain allowing bid opening on a private blockchain so that the bids can only be learnt by the auctioneers. Frauentaler et al. [44] introduced a scheme to reduce the operation cost of Ethereum-based blockchains up to 92% by employing a validation-on-demand pattern consisting of economic incentives. Yang et al. [45] proposed a technique that combines the miner-weighted history information with calculation difficulty to reduce attack problems. Malik et al. [46] proposed TrustChain which is a trust management framework consisting of three layers and helps blockchain in the tracking of supply chain participants. Wang et al. [47] presented ChainSplitter which is a hierarchical blockchain storage structure where a large amount of blockchain can be stored in clouds. Zheng et al. [48] developed a platform called NutBaaS which is a supply blockchain service over cloud computing such as system monitoring and network deployment. Xu et al. [49] proposed BlendMAS which is a blockchain-enabled architecture to secure data access control in an SPS system. Wang et al. [50] presented ArtChain, which is an integrated trading system based on blockchain.

SMART CONTRACTS FOR INDUSTRY 4.0

Blockchains can also be helpful for the automation of industrial processes, and it consists of multiple companies. We define a smart contract as it is a computer programme that helps in executing agreements that are established in between two parties and because there are some certain actions to happen when some specific conditions are met. Therefore, a smart contract executes a clause automatically when some conditions occur that have been programmed. The conditions depend upon some external services that take data from the real world and store it in the blockchain or vice versa [51]. On the basis of the collected information and its interaction with the real word, oracle has different types such as hardware oracle, software oracle, inbound oracle, outbound oracle and consensus-based oracle [52]. Information such as price, traceability of truck position, etc. can be handled by software oracle. The data are mainly collected from the web and then pushed into smart contracts. Hardware oracle extracts the data from the physical world directly. Inbound oracle fills information into the blockchain received from external sources that do not interact with the blockchain [53]. While outbound oracle also allows smart contracts to send processed data to the outside world, consensus-based oracle helps in combining different oracles so that they can compute the result of an event. Mei et al. [54] proposed a platform of stored value cards in which a smart contract expresses the service contracts. Dragoni et al. [55] demonstrated SC 2 (Secure Communication over Smart Cards), a system developed to address a key issue of the S × C framework.

BLOCKCHAIN-BASED INDUSTRY 4.0 APPLICATIONS

The use of blockchain benefits Industry 4.0 technologies a lot, but the application is also challenging. It may bring many features but the implementation of blockchain technology in the nodes is also very difficult. For example, by using blockchain, the security of data and communication can be increased but the privacy and integration of data will remain a challenge [56, 57]. Thus, below there is a description of blockchain-based Industry 4.0 developments for the clarification of the challenges and advantages.

IIoT

IIoT stands for the industrial Internet of Things as it concerns the traditional Internet of Things in an industrial setting. It mainly deals with interconnected sensors, devices, etc. that are networked together with industrial applications of a computer that includes the management of energy and manufacture. The IIoT can use the help of blockchain technology for performing the exchange of data and transactions during the different stages of processing where the transactions are signed and time-stamped. Various other benefits that blockchain provides are also the increased security of the IIoT, giving access to IIoT data in public or semi-public scenarios, it also guarantees data access and provides a communication medium.

ICPS

ICPS stands for interdisciplinary cyber physical system or cyber physical production system (CPPS) and it is basically a mechanism that is monitored or controlled by some algorithms that are computer-based. It collects the data, processes them and then stores them to control physical processes. These are basically distributed physically, and some can also be found on the internet; it points out the nature of an ICPS is decentralised. Due to this it is highly suitable for blockchain. In some industries the blockchain can also serve as the backbone of an ICPS.

Autonomous Robots and Vehicles

Autonomous robots and vehicles are the key to Industry 4.0 applications and it also needs to be performed through robots, cobots or autonomous ground vehicles. The difference between cobots and robots is that cobots can collaborate with human beings for certain tasks while a robot will only perform it in an autonomous way. Their main functionality is to search and transport things through a vehicle or factory. Unmanned aerial vehicles (UAVs) have also become very popular in recent years and they have been used in multiple industrial environments. The way blockchain can help automation is by allowing them to interact with each other in the network through smart contracts.

Cybersecurity

The protection of the system in Industry 4.0 applications is necessary. The industry has been sensitive to attacks in recent years so security is essential. Hash algorithms and secure public-key cryptosystems are used for blockchain technology. Some private blockchains can restrict the access for users, and it helps in reducing the numbers of possible attackers. The decentralised nature also plays a vital role in security since if one is under attack, then access to the information can be gained through the other nodes, therefore availability of data is guaranteed. Blockchain systems are prone to some attacks like sybil attacks which can disturb the behavior of system.

Simulation Software

The information collected through Industry 4.0 can be useful in many ways, such as to model the entity's behavior in the production system. It is also useful to determine the real world's current state. Blockchain can help simulation software in many ways, such as it can collect the data from various sources in a distributed system, and the improvement in availability is caused by different nodes that provide redundant information. It also verifies the authenticity of data and removes the uncertain parts, resulting in better accuracy. It is also helpful in distributing the tasks and calculations among various different nodes.

CONCLUSION

The IoT connects devices and gadgets with each other over the network. It makes our daily lives more efficient and less time-consuming or expensive. It also eases other aspects. Since it is touching every area of life, it thus makes it more vulnerable for attacks as several attackers try to take advantage of it. The IoT has played a major role in the development of Industry 4.0. Industry 4.0 has been prone to cyberthreats in recent years. Cyberthreats have affected factories financially and technically on a large scale. Security can also be enhanced using blockchain technology. In this chapter we have covered the security architecture for the IoT, how Industry 4.0 makes use of the IoT and the various security threats. The work explores the need for blockchain technology and smart contracts in Industry 4.0 and reviews the related application scenarios.

REFERENCES

1. Nakagawa, E. Y., Antonino, P. O., Schnicke, F., Capilla, R., Kuhn, T. & Liggesmeyer, P. (2021). Industry 4.0 reference architectures: State of the art and future trends. *Computers and Industrial Engineering, 156*. doi: 10.1016/j.cie.2021.107241, PubMed: 107241
2. Ghobakhloo, M., Fathi, M., Iranmanesh, M., Maroufkhani, P. & Morales, M. E. (2021). Industry 4.0 ten years on: A bibliometric and systematic review of concepts, sustainability value drivers, and success determinants. *Journal of Cleaner Production, 302*. doi: 10.1016/j.jclepro.2021.127052, PubMed: 127052
3. Ali, H. & El-Medany, W. (2019). IoT security: A review of Cybersecurity architecture and layers. In *2nd Smart Cities Symposium (SCS'19)*.
4. Böhm, M., Schiering, I. & Wermser, D. (2017). Security of IoT cloud services - A USER-ORIENTED test approach. *Advances in Cybersecurity, 2017*, 115–130. doi: 10.18690/978-961-286-114-8.10
5. Meneghello, F., Calore, M., Zucchetto, D., Polese, M. & Zanella, A. (2019). IoT: Internet of threats? A survey of practical security vulnerabilities in real IoT devices. *IEEE Internet of Things Journal, 6*(5), 8182–8201. doi: 10.1109/jIoT.2019.2935189
6. Tsiknas, K., Taketzis, D., Demertzis, K. & Skianis, C. (2021). Cyber threats to industrial IoT: A survey on attacks and countermeasures. *IoT, 2*(1), 163–186. doi: 10.3390/IoT2010009
7. Taneja, B. & Bhushan, B. (2021). Information and data security model: Background, risks, and challenges. *Emerging Technologies in Data Mining and Information Security*, 859–869. doi: 10.1007/978-981-15-9774-9_78
8. Bhayo, J., Hameed, S. & Shah, S. A. (2020). An efficient counter-based ddos attack detection framework leveraging software defined IoT (sd-IoT). *IEEE Access, 8*, 221631. doi: 10.1109/access.2020.3043082, PubMed: 221612
9. Sinha, P., Rai, A. K. & Bhushan, B. (2019). Information Security threats and attacks with conceivable counteraction. In *2nd International Conference on Intelligent Computing, Instrumentation and Control Technologies (ICICICT)*.
10. Saini, H., Bhushan, B., Arora, A. & Kaur, A. (2019). Security vulnerabilities in Information communication technology: Blockchain to the rescue (A survey on Blockchain Technology). In *2nd International Conference on Intelligent Computing, Instrumentation and Control Technologies (ICICICT)*.
11. Kuzlu, M., Fair, C. & Guler, O. (2021). Role of artificial intelligence in the internet of things (IoT) cybersecurity. *Discover Internet of Things, 1*(1). doi: 10.1007/s43926-020-00001-4
12. Igor, H., Bohuslava, J., Martin, J. & Martin, N. (2014). Application of neural networks in computer security. *Procedia Engineering, 69*, 1209–1215. doi: 10.1016/j.proeng.2014.03.111
13. Liao, Y. & Vemuri, V. (2002). Use of k-nearest Neighbor classifier for intrusion detection. *Computers and Security, 21*(5), 439–448. doi: 10.1016/s0167-4048(02)00514-x
14. Vidhya, M. (2013). Efficient classification OF portscan attacks using support vector machine In *International Conference on Green High Performance Computing (ICGHPC)*.

15. Althobaiti, O. S. & Dohler, M. (2020). Cybersecurity challenges associated with the internet of things in a post-quantum world. *IEEE Access*, *8*, 157381. doi: 10.1109/access.2020.3019345, PubMed: 157356

16. Khalid, A., McCarthy, S., O'Neill, M. & Liu, W. (2019). Lattice-based cryptography for IoT in a quantum world: Are we ready? In *8th International Workshop on Advances in Sensors and Interfaces (IWASI)*. IEEE Publications.

17. Singh, R. V., Bhushan, B. & Tyagi, A. (2021). Deep learning framework for cybersecurity: Framework, applications, and future research trends. *Advances in Intelligent Systems and Computing*, 837–847. doi: 10.1007/978-981-33-4367-2_80

18. Babbar, G. & Bhushan, B. (2020). Framework and methodological solutions for cyber security in Industry 4.0. *SSRN Electronic Journal*. doi: 10.2139/ssrn.3601513

19. Sethi, R., Bhushan, B., Sharma, N., Kumar, R. & Kaushik, I. (2020). Applicability of industrial IoT in diversified sectors: Evolution, applications and challenges. *Studies in Big Data Multimedia Technologies in the Internet of Things Environment*, 45–67. doi: 10.1007/978-981-15-7965-3_4

20. Khaleeque, R. & Mansoor, H. (2020). Internet of things (IoT) and it's needs. *Global Sci-Tech*, *12*(1), 38. doi: 10.5958/2455-7110.2020.00006.3

21. Quilter, J. D. (2014). *Smart security: Practices that increase business profits*, *1*, doi: 10.1016/b978-0-12-801515-5.00001-1

22. Robertson, P. W. (2020). Using supply chain analytics to enhance supply chain strategy processes. *Supply Chain Analytics*, 74–98. doi: 10.4324/9781003084020-4

23. Minchev, Z. & Gaydarski, I. (2020). Cyber risks, threats & security measures associated with COVID-19. *CSDM Views*, *37*(2020). doi: 10.11610/views.0037

24. Data protection-data security-privacy (1984). *Computers and Security*, *3*(1), 57–58. doi: 10.1016/0167-4048(84)90031-2

25. Vempati, A. S., Kamel, M., Stilinovic, N., Zhang, Q., Reusser, D., Sa, I., ... Beardsley, P. (2018). PaintCopter: An autonomous UAV for spray painting on three-dimensional surfaces. *IEEE Robotics and Automation Magazine*, *3*(4), 2862–2869.

26. Kyung-ho, S., Byuk-ik, K. & Tae-jin, L. (2020) Cyber-attack group analysis method based on association of cyber-attack information. *KSII Transactions on Internet and Information Systems*, *14*(1). doi: 10.3837/tiis.2020.01.015

27. Skopik, F., Bleier, T. & Fiedler, R. (2015). *Cyber Attack Information System: Gesamtansatz*. Xpert Press, 53–69. doi: 10.1007/978-3-662-44306-4_3

28. Bhushan, B. & Sahoo, G. (2017). Recent advances in attacks, technical challenges, vulnerabilities and their countermeasures in wireless sensor networks. *Wireless Personal Communications*, *98*(2), 2037–2077. doi: 10.1007/s11277-017-4962-0

29. Thomson, J. R. (2015). Cyber security, cyber-attack and cyber-espionage. *High Integrity Systems and Safety Management in Hazardous Industries*, 45–53. doi: 10.1016/b978-0-12-801996-2.00003-9

30. Cisco (2021, February 19). Cyber attack – What are common cyberthreats? Retrieved from https://www.cisco.com/c/en/us/products/security/common-cyberattacks.html. Cisco.

31. Rizov, V. (2018). Information sharing for cyber threats. Information & security: An [International journal], *39*(1), 43–50. doi: 10.11610/isij.3904

32. Alves, T. & Morris, T. (2018). Hardware-based cyber threats. In *Proceedings of the 4th international conference on Information Systems Security and Privacy*.

33. Cyber Security 2020 Keynote and industry panel speakers (2020). In *International Conference on Cyber Security and Protection of Digital Services (Cyber Security)*. doi: 10.1109/cybersecurity49315.2020.9138877

34. Larriva-Novo, X. A., Vega-Barbas, M., Villagrá, V. A. & Sanz Rodrigo, M. (2020). Evaluation of cybersecurity data set characteristics for their applicability to neural

networks algorithms detecting cybersecurity anomalies. *IEEE Access*, *8*, 9005–9014. doi: 10.1109/ACCESS.2019.2963407

35. Bhushan, B., Sahoo, C., Sinha, P. & Khamparia, A. (2020). Unification of Blockchain and Internet of Things (BIoT): Requirements, working model, challenges and future directions. *Wireless Networks*. doi: 10.1007/s11276-020-02445-6

36. Muzammal, M., Qu, Q. & Nasrulin, B. (2019). Renovating blockchain with distributed databases: An open source system. *Future Generation Computer Systems*, *90*, 105–117. doi: 10.1016/j.future.2018.07.042

37. Huang, K., Zhang, X., Mu, Y., Wang, X., Yang, G., Du, X., . . . Guizani, M. (2019). Building redactable consortium blockchain for industrial Internet-of-things. *IEEE Transactions on Industrial Informatics*, *15*(6), 3670–3679. doi: 10.1109/tii.2019.2901011

38. Morkunas, V. J., Paschen, J. & Boon, E. (2019). How blockchain technologies impact your business model. *Business Horizons*, *62*(3), 295–306. doi: 10.1016/j.bushor.2019.01.009

39. Althobaiti, O. S. & Dohler, M. (2020). Cybersecurity challenges associated with the internet of things in a post-quantum world. *IEEE Access*, *8*, 157381. doi: 10.1109/access.2020.3019345, PubMed: 157356

40. Dai, S. (2021). Quantum cryptanalysis on a MULTIVARIATE cryptosystem based on CLIPPED Hopfield neural network. *IEEE Transactions on Neural Networks and Learning Systems*, 1–5. doi: 10.1109/tnnls.2021.3059434

41. Xin, Y., Kong, L., Liu, Z., Chen, Y., Li, Y., Zhu, H., … Wang, C. (2018). Machine learning and deep learning methods for cybersecurity. *IEEE Access*, *6*, 35365–35381. doi: 10.1109/ACCESS.2018.2836950

42. Azwar, H., Murtaz, M., Siddique, M. & Rehman, S. (2018). Intrusion detection in secure network for cybersecurity systems using machine learning and data mining. In *5th International Conference on Engineering Technologies and Applied Sciences (ICETAS)*, pp. 1–9. IEEE Publications.

43. Desai, H., Kantarcioglu, M. & Kagal, L. (2019). A hybrid blockchain architecture for privacy-enabled and accountable auctions. In *IEEE International Conference on Blockchain (Blockchain)*, pp. 34–43.

44. Frauenthaler, P., Sigwart, M., Spanring, C., Sober, M. & Schulte, S. (2020). ETH relay: A cost-efficient relay for ethereum-based blockchains. In *IEEE International Conference on Blockchain (Blockchain)*, pp. 204–213.

45. Yang, X., Chen, Y. & Chen, X. (2019). Effective scheme against 51% attack on proof-of-work blockchain with history weighted information. In *IEEE International Conference on Blockchain (Blockchain)*, pp. 261–265.

46. Malik, S., Dedeoglu, V., Kanhere, S. S. & Jurdak, R. (2019). TrustChain: Trust management in blockchain and IoT supported supply chains. In *IEEE International Conference on Blockchain (Blockchain)*.

47. Wang, G., Shi, Z., Nixon, M. & Han, S. (2019). ChainSplitter: Towards blockchain-based industrial IoT architecture for supporting hierarchical storage. In *IEEE International Conference on Blockchain (Blockchain)*, pp. 166–175.

48. Zheng, W., Zheng, Z., Chen, X., Dai, K., Li, P. & Chen, R. (2019). NutBaaS: A blockchain-as-a-service platform. *IEEE Access*, *7*, 134422–134433. doi: 10.1109/ACCESS.2019.2941905.

49. Xu, R., Nikouei, S. Y., Chen, Y., Blasch, E. & Aved, A. (2019). BlendMAS: A Blockchain-enabled decentralized microservices architecture for smart public safety. In *IEEE International Conference on Blockchain (Blockchain)*, pp. 564–571.

50. Wang, Z., Yang, L., Wang, Q., Liu, D., Xu, Z. & Liu, S. (2019). ArtChain: Blockchain-enabled platform for Art Marketplace. In *IEEE International Conference on Blockchain (Blockchain)*, pp. 447–454.

51. Thanh, V., Rohit, S., Raghvendra, K., Le Hoang, S., Thai, P. B., Dieu, T. B., ... Le, T. (2020). Crime rate detection using social media of different crime locations and twitter part-of-speech tagger with brown clustering. *Journal of Intelligence and Fuzzy System*, 4287–4299.
52. Nguyen, P. T., Ha, D. H., Avand, M., Jaafari, A., Nguyen, H. D., Al-Ansari, N., ... Pham, B. T. (2020). Soft computing ensemble models based on logistic regression for groundwater potential mapping. *Applied Sciences*, *10*(7), 2469.
53. Jha, S., Kumar, R., Hoang Son, L., Abdel-Basset, M., Priyadarshini, I., Sharma, R. & Viet Long, H. (2019). Deep learning approach for software maintainability metrics prediction. *IEEE Access*, *7*, 61840–61855.
54. Mei, X., Ashraf, I., Jiang, B. & Chan, W. K. (2019). A fuzz testing service for assuring smart contracts. In *IEEE 19th International Conference on Software Quality, Reliability and Security Companion (QRS-C)*. doi: 10.1109/qrs-c.2019.00116
55. Dragoni, N., Lostal, E. & Papini, D. (2011) (SC)²: A system to secure off-card contract-policy matching in security-by-contract for open multi-application smart cards. In *IEEE International Symposium on Policies for Distributed Systems and Networks*, pp. 186–187.
56. Sharma, S., Kumar, R., Das Adhikari, J., Mohapatra, M., Sharma, R., Priyadarshini, I. & Le, D. N. (2020). Global Forecasting Confirmed and Fatal Cases of COVID-19 Outbreak Using Autoregressive Integrated Moving Average Model, ront. *Public Health*. doi: 10.3389/fpubh.2020.580327.
57. Malik, P. et al. (2021). *Industrial Internet of Things and Its Applications in Industry 4.0: State-of the Art, Computer Communication* (Vol. 166, pp. 125–139). Elsevier.

6 A Review of Precision Agriculture Based on Embedded System Applications

S. Ramesh and M. Vinoth Kumar

CONTENTS

INTRODUCTION

According to the report by United Nations world population will increase to 11 billion in 2100. All those people will need food. In today's world farmers face many issues: a growth of worldwide demand for food, a restriction in fossil fuels and water supply and environmental conditions. At the same time the farming of the future wants to be more productive and environmental impacts to be reduced. A new farming system is made of different applications and it consists of a farm with sensor deployment, to monitor and control automatic farms, information processing and network capabilities. Due to the rapid population growth and the main turns in the political, social and economic systems in India there is an urgent need of new development and to replace the existing agricultural conditions on the ground.

DOI: 10.1201/9781003097518-6

The stability and sustainability of Indian agriculture comes into question because of the pressure of demographics on India's natural resources. In the next decade the agriculture industry will be more important than ever before. In 2050, it will be necessary for global food production to have increased by 70% compared to current production, due to the growing population, as per the report of United Nations Food and Agriculture Organization. To increase food production and for easy access for farmers, agricultural companies are now turning to the Internet of Things (IoT) concept to meet these demands. In the farming sector new technology innovation begins right now. First, gas-powered tractors and chemical fertilisers were introduced in the 1800s. In the 1900s farmers started using satellites to organise their work. Here the proposed IoT concept will influence farming systems for the future. Among farmers, the concept of smart agriculture has become popular and high-tech farming has also become standard and this is because of agricultural sensors and drones. With the help of smart farming techniques, it is possible to increase crop productivity and farmers are about to start this concept in some parts of the country. Some work based on sensors is already installed in fields to monitor environmental conditions such as water irrigation, the temperature of the soil and acidity and at the same time climate change as well. In rural areas the above methodology is not implemented. So, our proposed concept is to implement the techniques in rural areas with our IoT reference architecture.

As an example, John Deere, one of the world's largest farming equipment manufacturers has already started to implement the smart concept with the help of internet connectivity, to monitor and display crop yields. The company is supplying self-driving tractors, similar to smart cars, which help the farmers to expand the productivity of the crop and free them up for other tasks.

Business Intelligence (BI) reported that IoT device installations in the farming sector were about 30 million in 2015 and are expected to be up to 75 million in 2020 (Figure 6.1). All over the world the U.S. leads in smart farming and has annual cereal production of 7,340 kg.

Now, all types of issues related to agricultural production will be managed by IoT applications. With help of IoT applications it is possible to increase the quantity and quality of the product as well as cost effectiveness in the production of agriculture. Today all types of farms are remotely monitored with the help of sensors and IoT applications. Smart agriculture and precision farming methods will take the future

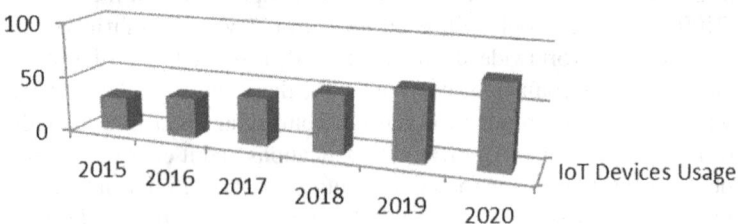

FIGURE 6.1 List of expected usage of IoT devices in the agri-farm sector are shown for the period from 2015 to 2010.

of farming to the next level. At the same time the above-mentioned technologies, are also able to predict current and future harvests. Moreover, the IoT-based applications will help the farmers in that they will have more time for other work.

BENEFITS OF THE PLATFORM FOR SMART AGRICULTURE

The world is getting smarter, but the scenario for farmers remains the same. To compare with other sectors, the technology involved in the farming sector is lagging. The purpose of building the IoT platform is to design the requirements of farmers' needs in the real world. With the help of the IoT platform the sector becomes smarter. It will monitor aspects like environmental conditions, pests, diseases, animal invasion and so on.

The ways in which the IoT platform can help:

- The data from the sensors are easily collected via cloud services with the existing system and connected equipment.
- The farmers will easily collect the data from remote locations and take the necessary actions.
- When compared to other applications, the IoT will speed up the process up to ten times.
- With the help of the IoT platform it will be able to access new market innovations and food supply chain management.
- Large amounts of data will be collected from the field, and it should be highly scalable and secure.

The next section discusses the literature survey, followed by a description of remote sensing techniques. The chapter then goes on to apply the intelligent systems, lists the possible applications of IoT in agriculture, ending with a discussion and conclusion.

LITERATURE SURVEY

The literature survey outlines that by applying the IoT applications in the field, farmers are able to increase crop productivity. Hence, embedded system applications as indirect tools can be exploited to uplift the current state of Indian agriculture.

This chapter is useful to researchers and those who are working in smart farming methods. The different authors explain their work on various problems occurring in the field. A study of the precision agriculture sector is shown in Figure 6.2.

REMOTE SENSING

Remote sensing is one of the oldest techniques in precision agriculture (PA) sector. Various methods of remote sensing techniques in PA are explained below.

Zhang et al. (2004) proposed the identification of disease in the tomato plant by spectral analysis [1]. In tomato plants, late blight is one of the severe diseases and it affects the productivity of the crop. An airborne visible and infrared imaging

Smart Field Study of Various techniques in Precision Agriculture Field

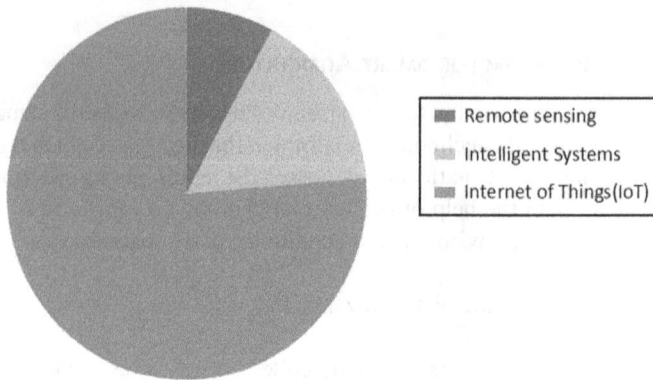

FIGURE 6.2   Surveys of various techniques in precision agriculture.

spectrometer (AVIRIS) is used in the plant to differentiate the healthy plants from infected ones, in the wavelength range of 400–2500 nm. Image analysis is divided to six indices DI1 = B1/N1, DI2 = B1/N2, DI3 = B1/N3, DI4 = B1/N1, DI5 = B1/N2, DI6 = B1/N3.A healthy index is denoted as DI1; average band values within 800–900 nm are denoted as B1; 1000 1100 nm denotes B2; average band values of N1,N2 and N3 are denoted as 600–650 nm,1400–1700 nm and 2000–2200 nm.DI1, DI3 and DI4 give the best results from the six indices. Disease comparison is made by four stages. Above 13 it is at a healthy stage; between 2 and 13 it may be diseased; at stage 4, the disease can be 5.2, at stage 3 the disease plants around 6, when it indicates less than 2, it corresponds to the soil. Finally, the results indicate that DI1 is suitable to separate the infected plants from healthy ones. The research is applied in California.

Huang et al. (2014) proposed new spectral indices (NSI) to identify different diseases in crops [2]. The diseases considered were aphids, yellow rust and powdery mildew in wheat plants. The RELIEF-F algorithm was used to extract the different diseases from leaf spectral data. The classification accuracy for infected and healthy leaves was 86.5%, 85.2% and 91.6, 93.5% respectively. The work took place in the month of April and a spectral region of 400–1000 nm was analysed. NSI uses two wavelengths as single and normalised wavelength differences. The advanced stage of different diseases is found by a single wavelength. The normalised wavelength differences quickly detect disease in the hyper spectral data. The best combination of wavelengths is done by the RELIEF-F algorithm. This methodology will efficiently identify plant diseases.

Mewes (2009) proposed band selection techniques applied on fungal infected and vital wheat stands, to differentiate and localise the infected area [3]. Some 84 sample points were taken, with image pixels of 3 × 3 and were divided into 70% for testing and 30% for training. They were classified as vital and infected. The vital class denotes the healthy wheat stands and infected class denotes the pathogen-infected plants. Five percent of diseases beyond the sampling point are classified as infected, the remainder as vital. Bhattacharyya distance (BD) is used to measure the separation between the two classes. Various supervised classifications are applied on the selected bands to obtain the evaluation purposes, such as decision tree (DT), spectral angle mapper (SAM) and maximum likelihood classification (MLC). These classifications are compared to get the calculation accuracy from the test data. Two separate techniques were tested. The BD performed better than DT algorithm. The BD performed well in accuracy and processing time.

INTELLIGENT SYSTEMS

Baranwal et al. (2016) designed the idea of a management information system (MIS) [4]. The MIS will improve the process of agricultural information and increase agricultural production. The system is divided into three layers: data transmission, data collection, information analysis and process. Agricultural production consists of various layers: crop storage and crop growth. Crop storage will be monitored by a sensor, GPS is used to locate the farmland and RFID is used for crop tracking. Finally, using this intelligent system will provide efficiency and reliability to agricultural production and enable large-scale industrial application and high-level management.

Ryu et al. (2015) designed the IoT methods which focus on smart farming systems for end users [6]. Automated farm systems are developed by using components such as sensors and controllers, an IoT gateway, monitoring and IoT service platform. Connected farm implementation is made by Mobius, &cube and physical devices. Mobius is an IoT service platform, &cube is a device software platform which can be installed into IoT gateways. Three types of sensors (humidity, temperature and CO_2) are deployed in connected farms to monitor the environmental conditions. All the sensors are wirelessly connected with Raspberry-Pi installed and Zigbee networks. Each sensor collects the data at 20 second intervals. Collected data can be operated remotely through the &cube. Mobius is an IoT application developed with RESTAPI. Finally, with the developed smart mobile application it will enable users to remotely monitor and control the connected farm.

Qiu et al. (2014) designed the concept of the environmental monitoring control system [7]. The system detects parameters like humidity, temperature and light intensity in real time using the Zigbee network. The system architecture consists of upper machine processors, an environment factor acquisition block, an intelligent control terminal block; the processor used is a S3C2440 microprocessor with the ARM920T core. The system software design is divided into three parts: S3C2440 microprocessor, Zigbee-based coordinator software and end-device software, SQLITE3 database. In real time the test result of the data sets the value for early warning and selects the node which has to display. A combination of hardware and software gives the test results of the system. Finally, the system is feasible in real time for an intelligent greenhouse environment.

Furriel et al. (2015) proposed measuring the soil compaction on croplands. Soil compaction information is provided by the intelligent apparatus of the cone index [8]. The proposed system will replace the current system by quality with higher performance. The intelligent system consists of designed penetrometer sensors data acquisition, data processing, calibration of the system and an intelligent embedded system. System usage will save in production and water irrigation. Water irrigation is done by the same soil compaction procedure. It also saves the loss of seeds and root growth. Power values of the system are done by electrical measurements. The new intelligent system proposed for soil compaction is without using any load cells and it provides good results with high resolution and interpretation data for the simple value analysis.

Yunping et al. (2009) proposed the framework for soil sampling by using an intelligent system [9]. For soil sampling five modules had been set based on global position technology and geographic information system (GIS). The five modules are design, navigation, location, data analysis and detection analysis. The software development environments are visual C++ and maps object mode in the Windows 2000 platform. Sampling path design is done by elastic algorithm. The system will optimise the sampling path for soil. The system is classified as part hardware and part software. The hardware part includes GPS and sensors; the software part consists of database and application software. In PA soil investigation is an important part but due to low efficiency, poor design and high cost it still does not meet the requirements. This system will replace all issues related to soil in the field. The agriculture sector needs an efficient, accurate, automated soil sampling information system. This will be done by the system.

INTERNET OF THINGS (IoT)

Mohd Kassim et al. (2014) proposed building a decision support system using wireless sensor networks [11]. Nowadays, decision support is increasing fast in the agriculture sector. The system will monitor and analyse the land and crops and provide real-time information to the farmers to make the right decision. The intelligent greenhouse monitoring system (IGMS) is designed to monitor environmental conditions. It will optimise agricultural production and design in a user-friendly manner with low cost. The results of the system indicate that when to compared to scheduled irrigation, automatic irrigation is better in the field. To optimise the usage of water and fertiliser, automatic irrigation will help with the moisture and health of the plants.

Harun et al. (2015) designed the decision support system for real-time field problems [12]. The IoT system includes hardware, network architecture and software control of the process. The sensors monitor the data in the feedback loop, and a threshold value will be calculated to control the devices. This is used to monitor the environmental conditions such as crop growth, production efficiency, improve product quality and conserve energy. The test shows that automatic irrigation performed better that scheduled irrigation. By applying this system on the field, irrigation water of 1500 litres can be saved and is able to supply this water to1000 trees per day.

Mat et al. (2016) proposed an IoT system for agriculture optimum irrigation to consist of a wireless moisture sensor network and a wireless sensor network (WSN) [13]. The system will monitor irrigation, because inefficient irrigation will lead to wastage of the water. A test was conducted to differentiate the two methods, scheduled irrigation and feedback irrigation. Scheduled irrigation will supply water to the plants in specific time periods. Feedback irrigation is to irrigate plants when the moisture or level of media wetness reaches a predefined value.

Konstantinos et al. (2007) proposed a wireless sensor network to measure the electrical conductivity in the field and it is different from conventional network implementation [14]. WSN topology is able to build with the decision of electrical conductivity. A lot of ad hoc routing algorithms will be found on a WSN, and each algorithm has its own framework. It is very difficult to compare all algorithms. So RMASE is used here to compare the algorithms. It is a routing algorithm and it consists of throughput, success loss rate, latency and energy consumption efficiency. Prowler is used to implement the RMASE. Prowler is a type of tiny OS. The combination of precision farming and WSN applications is an existing new era of research that will increase agricultural production and improves water management. Using this system will reduce the cost and it is suitable for all kinds of fields and cultivation.

Li et al. (2011) designed the precision agriculture monitor system (PAMS) to monitor environmental conditions and the crop growth period [15]. The system is classified into two parts: remote control and environmental monitor. The NPU mote is a type of sensor and is deployed to sense the environment. The PAMS is developed with ATMEGA 12L 8-bit microcontrollers with WSN, communication server and gateways. The deployed node will sense the information and send it to the server via gateways. The remote user consists of GUI and is used to monitor the data collected in the field. The system will help for long-time deployment and energy efficiency. The system will satisfy farmers because it will control and manage the environmental data in real time.

Hu et al. (2010) proposed the wireless sensor and actor networks (WSAN) method which is used to collect the relevant data in the field [16]. The method proposes the system architecture and analyses the requirements. The hardware and software implementation with individual network components includes sensor node, actor node, video nodes and the gateway. Finally, it analyses network performance using various nodes. All the process will be controlled and managed by the control centre. Overall, the network performance results obtained from the test is 9.1 s.

Gayatri et al. (2015) proposed how the IoT is applied to Indian agriculture systems [17]. The major characteristics of emerging technologies such as the IoT and web services are utilised to construct an efficient approach to handle the enormous data involved in agrarian output. The approach uses the combination of the IoT and cloud computing that promotes the quick process of agricultural extension, helps to perform smart agriculture solutions and systematically answers the problems related to farmers. The author explains about various sensors applied to the agricultural process and near and far communication nodes are also explained. It is concluded that the proposed system is applied to Indian agriculture for the needs of farmers' access.

Ramundo et al. (2016) proposes IoT technology for food supply chain management [18]. The food sector is challenged due to increased population. Food supplied to the consumers should be safe and good quality and also provide an awareness of the products like food quantity, quality and expiry; all are explained in this sector. Concepts derived from the supply chain are farm production, processing, packing, sales and marketing, logistics, consumers and waste. All the processes will be carried out successfully by IoT applications. Still, in rural areas they are not aware of the food supply chain due to the low telecommunication range; the inadequate communication system is to be solved by this system.

Vatari et al. (2016) proposed monitoring and analysing the many crops grown in a greenhouse environment [19]. It is designed with a sensor network and is used to monitor the temperature, humidity and soil controls. An IoT device is used to monitor the environmental data from anytime and anywhere, and data are stored in the cloud. The IoT and cloud computing will combine to make the greenhouse effective. The system will sense the data and information will be sent to the user via the cloud and they will take necessary action on the data. Finally, the results of this method will enable farmers to achieve a high crop yield, better quality, prolonged production period and less use of chemicals in the field.

Khattab et al. (2016) designed the IoT architecture to predict environmental conditions in the field [20]. The architecture is designed with three layers: collection of the required data and analysis, the front-end layer will collect the environment information, data storage and processing takes place in the back-end layers. The performance tests of system are carried out by the proposed architecture. Using this system three categories of environmental data are measured: wind speed, rain and air temperature, humidity. The system facilitates the collection of data on environment conditions in real time.

Sreekantha et al. (2017) proposed an IoT system to provide the solutions for farmers' problems occurring in the field [21]. The IoT would fetch details about crop growth, pest detection, animal invasion in the field and weather. A sensor will gather the data and transmit to the device for farmers' access. A WSN is used to monitor the farms and microcontrollers are used to automate the farm process. An image the farm can be viewed using a wireless camera. With help of the IoT device the farmers can monitor their farm anytime and from anywhere. Using this proposed technology, the farmers will be able to increase their productivity and profit.

Carrasquilla-Batista et al. (2016) proposed a hardware- and software-based electronic system designed to automatically find the over-drain measurements in greenhouse horticulture using the IoT. Kalman filtering is applied for over-drain measurements [22]. The liquid flow meter accuracy is +/−10% and after using a Kalman filter it is reduced to +/−1% accuracy. Finally, the results reveal that the system proposed in this chapter used to generate a module of crop growth and secondary variables is used to analyse the plant growing process and increase the production of vegetables while minimising input of water, nutrients and fertilisers.

Hu et al. (2011) designed the embedding crop growth models (CGMs) implemented into an IoT application to make the system more intelligent and adaptive [23]. The system is developed in a greenhouse environment with essential data like temperature, humidity and light intensity collected by sensor. After the data is processed, the relevant algorithms will give output to the management system about the crop growth condition.

Zhao et al. (2010) proposed a remote monitoring system consisting of wireless communications and the internet [24]. Data collected in real time on temperature, humidity and soil signals, are transmitted to an M2M support platform through WAP or SMS, and the end user can analyse the information to support production. This is the best example of IoT applications in agriculture to obtain the relevant data in the real time. With this system greenhouse site monitoring is done by mobile wireless communications. The system will provide the output in high performance which improves the efficiency. The system is able to revise the control of environmental parameters.

Liu et al. (2014) designed the IoT architecture based on Internet of Things naming service (IoTNS) [25]. With the help of IoTNS it is able to enhance the access to productive and to reduce the complexity of the object name service (ONS). The implementation of food supply chain management for agricultural products is done by IoTNS. With the help of food supply chains, it is able to predict the farm production, expiry of the product, packing, consumer awareness and waste. Finally, the IoTNS system will give the perfect output for farmers and consumers to provide the proper information about agricultural production and products in the market; it will analyse and give information about market needs as well.

Bing et al. (2012) designed an intelligent system for the development from current to modern agriculture methods [26]. The system is applied to fruit production and organic melons. The system is made up of existing technologies such as sensors and RFID and designed with three platforms including the expert systems service platform, the intelligent production management platform and the integrated trading platform. In platform one, a mathematical model is used as the setup to obtain the data from the growing melons and a decision will be taken. The second platform is used for environmental control, fertiliser and water supply. The third platform is designed with a traceability function to optimise fruit planning management. The system will save the farmers time and increase the quantity and quality of the fruits.

Ye et al. (2013) developed the precision agriculture management system (PAMS) with the help of the IoT and web GIS applications [28]. It is used in China on an organic form, to increase the shortcomings of the current development of precision agriculture. PAMS includes four architectures: the IoT infrastructure platform, the spatial information infrastructure platform, the agriculture management platform and the mobile client. Users can monitor and manage agricultural production using PAMS. The design of PAMS includes information infrastructure, database establishment, web GIS module, local system and mobile client. It will promote farm efficiency to the next level of management. PAMS is developed with hardware and software components and involves multiple systems and processes.

Tuli et al. (2014) proposed the 'agri-assistance' cloud deployment model which provides information about financial and agriculture issues to Indian farmers in rural areas [29]. This e-agriculture concept is taken as a base for the agri-assistant and the cloud complexity concept is used for the purpose of traditional services and is provided by Government of India. The components of agri-assistant consist of an administration and maintenance system, information storage system, government repositories and farmers' interface devices. The ICT is suitable for Indian agriculture's current development and in the future it will be included with technologies like the IoT. This model not only focuses on exploiting the technological advancement but at the same time makes sure that no financial burden is added on poor farmers by making use of the cheapest available resources and existing services already being provided by the Government of India.

Shi et al. (2015) proposed that IoT technology is used to monitor the diseases and pests in the field. The monitoring system will collect the data via sensors and data will be processed [30]. All kinds of information in the field related to agricultural pests and diseases will be monitored by this system. The disease and pest control system is divided in to three systems and three levels: data acquisition equipment in the IoT, architecture of the IoT in agricultural monitoring systems and a disease and insect pest monitoring platform. The system will monitor the diseases and pests which will result in high yield and high quality on the field.

Zhang et al. (2011) proposed agricultural environmental monitoring based on ARM7 for a real-time image acquisition system [31]. To answer the issues in an agricultural environment for image acquisition, motion capture and real-time video acquisition, a 32-bit embedded processor (ARM S3C44BOX) is used. The main purpose of the processor system is low power consumption, low cost, multi-functionality and good expansion. The combination of system is made by sensors and a GPS module to monitor the environmental conditions. The system includes hardware and software modules. The hardware module design of the S3C44BOX processor consists of control signal analysis, input port settings, clock settings and the choice of work modes size. Software design includes boot loader transplantation and uclinux transplantation. S3 Processors are able to expand the CF card interface, FLASH and SDRAM. Finally, the result states that the system is used for the agricultural environment and can be further expanded to develop with the help of wireless transmission. The system is able to realise the functions like image and video acquisition for movement detection and storage. Video acquisition is implemented by using the new design of the 32-bit ARM processor.

Mohanraj et al.'s (2016) proposal to monitor and control the agricultural field by IoT application is created on the basis of an e-agriculture framework with monitoring modules and KM-knowledge base [32]. With the concept of ICT it changes the path of rural farmers and can replace existing methods in Indian agriculture. The system design and realisation should consist of knowledge base and monitoring interface, realised inputs, knowledge acquisition and knowledge flow. The monitoring interface includes hardware and software requirements and system architecture. The hardware requirement consists of a TICC3200 LaunchPad and Arduino UNO board. The software requirements consist of Energia MT. It is an open-source platform to compile and execute the embedded Ccode. The system architecture module of the monitoring interface comprises the application of its main ability and functions. The knowledge base of the user interface is used for the decision-making system. The structure of the knowledge base consists of a variety of crop information like geospatial data, various inputs like market availability, weather prediction and knowledge acquisition. The monitoring modules contain information like problem identification, various stages of plant growth, reminders and an irrigation planner. With the help of the devised algorithm, the evaporation method will calculate the water need for plants on a daily basis. Finally, the results of the system state it is suitable for the Indian agricultural sector.

Markovic et al. (2015) proposed IoT applications to monitor and control agricultural production [33]. It is designed with the Arduino platform, sensors and some library function. Sensors will collect the data and send signal information to the actuators to control the process. The system connects to the internet and the user can remotely monitor the production process and condition. An IoT-based monitoring and control system is used on the farms. The advantages of the system will include a reduction in costs and the facility to trace farm production.

Chen et al. (2012) designed the framework of digital agriculture by IoT applications [34]. The architecture of the system includes radiofrequency identification technology, electronic product code technology, Zigbee technology. Using Electronic Product Code (EPC) technology, the problems in the agriculture sector can be solved. As field information is obtained primarily through manual measuring experiments or judgements in traditional agriculture it takes a lot of labor power and the data accuracy is low. This system will solve all types of issues in the field.

Patil et al. (2016) proposed a system consisting of a combination of internet, wireless communication and a remote monitoring system (RMS) [35]. The objective of the work is to collect real-time data in agriculture such as weather patterns and crops through SMS.A camera module is also used to monitor the crop and it will be controlled by the decision. Farm fields consist of different crop areas. In the crop areas Ubimotes are installed. Data from UBI sense motes will be transferred to a UBI mote server module. A decision support system will be implemented for alerts for crop monitoring; the client-side module consists of a web application as Android OS mobile application. Finally, the system is able to monitor the crops from anywhere and at anytime.

Lee et al. (2013) proposed an IoT system for agricultural production. The agricultural production system will increase the stability of the food supply chain and improve the demand for agricultural products [36]. The IoT monitoring system will analyse the crop environment and the decision-making system will analyse the future harvest statistics. IoT technology is implemented in GUI visualisation software. The system will monitor field activities like the crop environment and the quality of the crop. Farmers can be free while applying this system in the field and also predict the crop lifecycle from sowing to selling. Finally, the system will give the information of current conditions int he field as well the future harvest.

Baranwal et al. (2016) proposed, designed, analysed and tested a system based on the IoT to transmit information to the user [37]. Remote users can monitor the crop field. The system is integrated with electronic devices, sensors and uses Python scripts. A test case achieved access of 84.8%. The intelligent security system has the ability to analyse the data and transmit information to a remote location. Devices processes include of analysis, transmission and data collection. The architecture layer consists of an application layer, perception layer and network layer. To monitor the remote location, an IP-based CCTV security camera is used with the network connectivity. RPI act as the server and will analyse the data and transmit it to grain stores. Finally, a system with security cameras is suitable for agricultural farms.

Li et al. (2014) proposed that an agricultural IoT field can be monitored and controlled by wireless sensor nodes. With this method, the end user could monitor the crops through the internet with the aid of browsers. The BIS mode includes the controlling system, multi-user and multi-site remote monitoring. The focus has been made on data transmission and wireless sensor modes control. The system is partitioned into three layers: GPRS module, server and user. The software implementation used socket technology and a JAVA multi-threaded mechanism. Finally, with the help of the system remote users can monitor the crop from anywhere and at anytime simply by having an internet browser.

PROPOSED SYSTEM

The proposed system architecture is described into two parts: smart field monitoring using a mobile app and real-time data from the farm are collected through sensors. First, we discuss the mobile app. The farmers in the field will capture the images with the mobile phone and verify those images with the database. Mobile devices will be designed with farmers' local languages. Captured images will be compared with database images and obtain information about them. If any new disease images occur it will be transferred to the remote user and they will get help with the diseases. This mobile device app will detect the diseases in the crop early and stop them spreading to other crops. Secondly the sensors are deployed in the field (humidity, sensors, CO_2) to monitor the environmental conditions. The system is made up of sensor connectivity, WSN and a remote user. The sensor will monitor the weather conditions and transfer the data to the remote user using the Zigbee network. Real-time data are collected at 20-secondintervals. The remote user will monitor the field from their location anywhere and at anytime. With the help of this system it is able increase the efficiency of the crop and quality as well. Applying this proposed system in the field, farmers will be freed from their work and the field also gets smarter. A brief description of the proposed architecture is shown in Figure 6.3.

SMART PRECISION AGRI-FARM USING PROPOSED IoT METHOD

COMPARISON OF VARIOUS METHODS IN THE PRECISION AGRICULTURE (PA) FIELD

See Tables 6.1–6.3.

DISCUSSION

Figure 6.3 shows the technique with different methods and an agriculture farm field in each research. The research paper also explains how the various techniques are applied in PA. Although embedded applications are used to monitor the parameters of pests, diseases, environmental conditions and so on, the development of future agriculture is lagging in some others. Now the current requirement in the agriculture sector is to increase the production for coming decades. The study of the paper has been made over the last 15 years. Some techniques lead to this discussion.

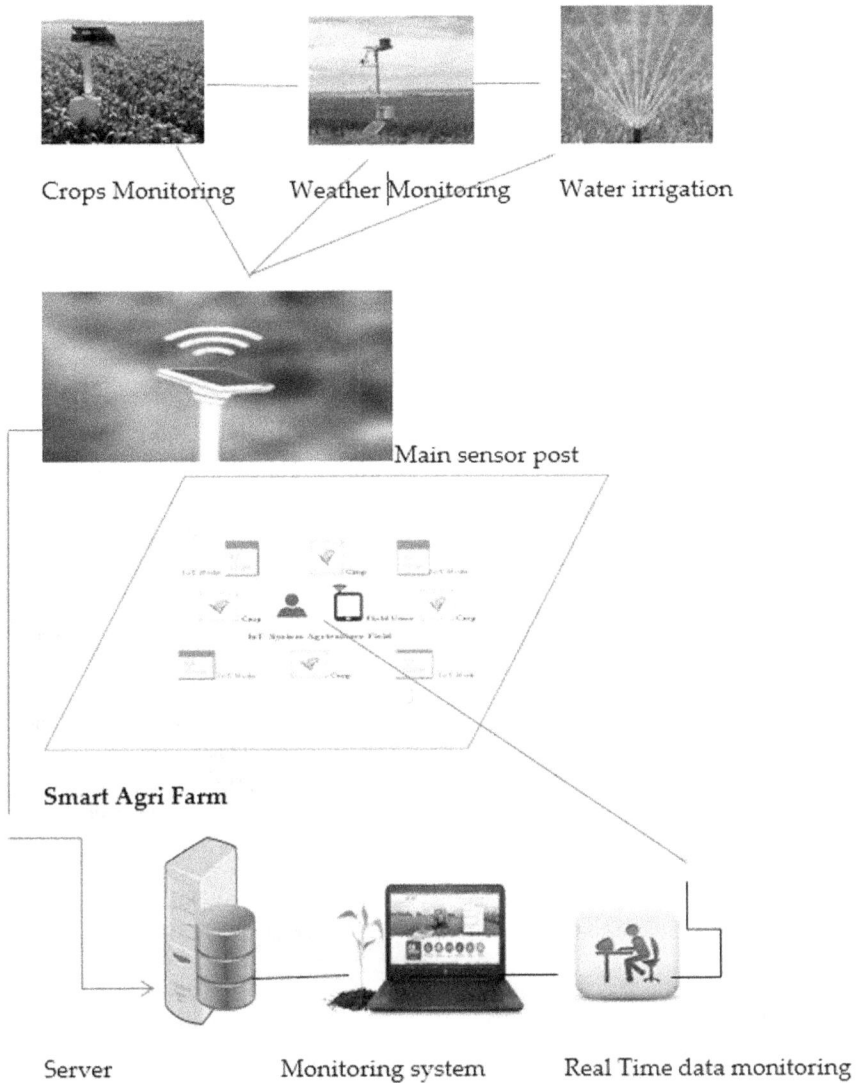

Crops Monitoring   Weather Monitoring   Water irrigation

Main sensor post

Smart Agri Farm

Server   Monitoring system   Real Time data monitoring

FIGURE 6.3   Proposed IoT architecture for an agri-farm field.

Performance Standard Is Too Accurate

Lots of methods used to expand the new techniques are collected under very exact conditions; remote sensing techniques are one of the regular operations and are ide-ally suitable in the early stage of research. This is applicable in real-world applica-tions. A large number of studies need get to the point of testing the procedure to deal with more realistic conditions because this greatly limits their scope.

TABLE 6.1

Remote Sensing Methods

S.No.	Reference	Parameters	Salient Features
1	Zhang et al. (2004)	Disease infections in tomatoes (late blight)	Out of six indices, D1 is suitable to separate the healthy plant from infected ones
2	Huang et al. (2014)	Powdery mildew, yellow rust and aphids	The accuracy obtained is 86.5%, 85.2%, 91.6% and 93.5%
3	Mewes et al. (2009)	Fungal-infected wheat	The BD algorithm is better than the DT algorithm

TABLE 6.2

Intelligent Systems

S.No.	Reference	Parameters	Salient Features
1	Yan-e et al. (2011)	Analysis of the features of agricultural production	MIS is suitable for agricultural production
2	Qiuet al. (2013) [5]	Intelligent monitoring platforms for facility agriculture ecosystem	The system is able to control the diseases and increases the production efficiency and production cycle
3	Ryuet al. (2015)	Used to monitor humidity, temperature and CO_2	IoT applications and mobile apps are used to monitor the field from a remote place.
4	Qiuet al. (2014)	Greenhouse environment factors to monitor humidity, temperature and light intensity	The system processes a feasible solution for intelligent greenhouse environment
5	Furrielet al. (2015)	Intelligent apparatus gives information about soil compaction	Apart from information about soil compaction, the system is used for weather, irrigation and the growth of roots.
6	Yunpinget al. (2009)	Intelligent system framework has been constructed for soil sampling	From the system it gives the result of efficient, accurate and automatic soil sampling information system

Shortage of Scientific Identification

The simplest solution is explained in the wireless network concept, but techniques like remote sensing, the IoT and an intelligent management system want to be more effective. In this case it is not effective.

Techniques Are Too Specific

The common techniques have been able to determine agricultural environmental monitoring. A lot of sensors had been suggested, not to mention that only one of the

TABLE 6.3

Internet of Things (IoT)

S. No.	Reference	Parameters	Salient Features
1	Kalaivani et al. (2011)	Soil fertility, growth of the crop, soil moisture content and temperature	Finally system is suitable to monitor the weather conditions
2	Mohd Kassim et al. (2014)	Farming resources optimisation and land monitoring	To increase the result of yield crop, water fertiliser optimisation is used
3	Harun et al. (2015)	Crop growth, production efficiency, improve product quality and conserve energy	Usage of water fertiliser, moisture level and health of the plant
4	Mat et al. (2016)	Wireless moisture sensor network is used for agriculture optimum irrigation	1500 ml per day of water can be saved and supply to 1000 trees per day.
5	Konstantinos et al. (2007)	WSN is used to measure the electrical conductivity in the fields	It is suitable in all parts of field and cultivation
6	Li et al. (2011)	PAMS is an intelligent system which can monitor the agriculture environment of the crops and to improve the output of the crops	PAMS will monitor the field from a remote location
7	Hu et al. (2010)	Wireless sensor and actor networks (WSAN) are used to monitor the change in environment	It predicts the environmental changes and also saves the field from animal invasion and pests
8	Gayatriet al. (2015)	IoT-based smart agriculture will solve the issues in the farmer's field	Sensors are deployed in the field and give information to the farmers' access
9	Ramundoet al. (2016)	State of the art and other emerging technologies framed in the whole supply chain management, and all technology bring to the sector	With help of this technology, the rural area problems are solved. like low telecommunication coverage and network connectivity
10	Vatariet al. (2016)	Monitor the crop growth, temperature, humidity and soil controls	This system will increase the efficiency of crop yield and reduce the chemicals in the field
11	Khattabet al. (2016)	Optimise the quality of the crops, minimise the negative environmental impacts	The proposed system is suitable for environmental data in real time
12	Sreekanthaet al. (2017)	Monitor the field like soil moisture, weather, pest detection, animal intrusion and crop growth	The farmers can monitor the field at anytime and from anywhere
13	Carrasquilla-Batistaet al. (2016)	To monitor the over-drain measurements in the field. Kalman filtering is applied to control the over-drain measurements	Before liquid flow meter in the field is +/−10% by using the Kalman filter it is reduced to +/−1%

(Continued)

TABLE 6.3 (CONTINUED)
Internet of Things (IoT)

S. No.	Reference	Parameters	Salient Features
14	Huet al. (2011)	CGMS model to monitor the humidity, temperature and light intensity tags	The system makes the farm more intelligent and adaptive
15	Zhao et al. (2010)	Monitor temperature, humidity and soil signals by wireless network support platform (M2M)	The system provides high performance, reliability and improves efficiency
16	Liu et al. (2014)	IoTNS is implemented for the supply chain management of agricultural products	The system provides information to the consumer about the product
17	Binget al. (2012)	Early detection of symptoms monitors organic melon and fruit production	The system finally improves the quality of the fruits
18	Jayaramanet al. (2015)	Monitor the real-time field environment. Do-it-yourself principle is used	Digital agriculture is used to monitor the environmental information
19	Yeet al. (2013)	To overcome the shortage of agriculture production, the system used a concept called PAMS	PAMS shows good results in agricultural production
20	Tuliet al. (2014)	The 'agri-assistance' model gives information about the latest technology in farmers' fields in rural areas	The system will also alert the poor farmers about the cheapest resources, it is provided by Government of India
21	Shiet al. (2015)	To monitor the diseases and pest control	The results of the system show in high yield
22	Zhang et al. (2011)	For real-time data acquisition of agricultural environmental monitoring the ARM7 is used	The system monitors the field by image and video acquisition, and it can be further extended with the help of wireless transmission
23	Mohanraj et al. (2016)	Automatic field monitoring and automation using IoT for environmental condition	The system success in various stage of research
24	Marković et al. (2015)	Monitoring and controlling of agricultural production	The system connects with internet, the farmers can remotely monitor the production process and condition
25	Chen et al. (2012)	Field-related problems are solved by EPC technology	By using this concept, a lot of labor power can be reduced, and data accuracy also becomes low
26	Patil et al. (2016)	Real-time data collection in the field such as weather patterns and crop growth are done by RMS	The relevant data in the field are monitoring by RMS
27	Lee et al. (2013)	Monitoring system to analyse the crop environment	This system will predict the current and future harvests
28	Baranwal et al. (2016)	The remote users monitor the field with an IoT system	84.8% of accuracy achieved by this test
29	Li et al. (2014)	Monitor the field by internet browser	The system with internet access is able to monitor the field from any location

sensors is properly explained, but those sensors' connectivity needs to be explained clearly with at least one of the environmental conditions. Many of the papers cannot to move to another stage and thus want to be more detailed. For more information on the subject readers can refer to the bibliography.

CONCLUSION

In the future, with the help of IoT developments the agriculture industry will become smarter and easier. A large amount of data could be collected and stored in a database and be easy to analyse the complex types of issues. Smart farming techniques are very useful for end-users to control (cellular phone) the farm from anywhere. With the help of this proposed system farmers can monitor their fields from anywhere. The system will monitor and control parameters like environmental monitoring, pests and disease control. When compared with the U.S. and China the IoT-based implementation of smart farming in our country has a long way to travel to obtain the benefits of smart farming techniques. Still, in our country, most of the farms are lagging in the smart farming concept because of issues like network connectivity and communication in rural areas. So, the proposed IoT-based architecture could solve the issues in rural areas very rapidly

REFERENCES

Baranwal, T. & Pateriya, P. K. (2016). Development of IoT based smart security and monitoring devices for agriculture. In *6th International Conference on Cloud System and Big Data Engineering (Confluence)*, Noida, p. 597602.

Bing, F. (2012). Research on the agriculture intelligent system based on IOT. In *International Conference on Image Analysis and Signal Processing*, Huangzhou, China, pp. 1–4.

Carrasquilla-Batista, A., Chacón-Rodríguez, A. & Solórzano-Quintana, M. (2016). Using IoT resources to enhance the accuracy of over drain measurements in greenhouse horticulture. In *36th Central American and Panama Convention (CONCAPAN XXXVI)*, San Jose, Costa Rica, pp. 1–5. IEEE Publications.

Chen, X.-Y. & Jin, Z.-G. (2012). Research on key technology and applications for Internet of things. *Physics Procedia, 33*, 561–566.

Furriel, G. P., Calixto, W. P., Alves, A. J., Campos, P. H. M. & Domingues, E. G. (2015). Intelligent system for measuring soil compaction on croplands. In *IEEE 15th International Conference on Environment and Electrical Engineering (EEEIC)*, Rome, pp. 1357–1361.

Gayatri, M. K., Jayasakthi, J. & Mala, G. S. A. (2015). Providing smart agricultural solutions to farmers for better yielding using IoT. In *IEEE Technological Innovation in ICT for Agriculture and Rural Development (TIAR)*, Chennai, pp. 40–43.

Harun, A. N., Kassim, M. R. M., Mat, I. & Ramli, S. S. (2015). Precision irrigation using wireless sensor network. In *International Conference on Smart Sensors and Application (ICSSA)*, Kuala Lumpur, pp. 71–75.

Hu, X. & Qian, S. (2011). IOT application system with crop growth models in facility agriculture. In *6th International Conference on Computer Sciences and Convergence Information Technology (ICCIT)*, Seogwipo, pp. 129–133.

Hu, J., Shen, L., Yang, Y. & Lv, R. (2010). Design and implementation of wireless sensor and actor network for precision agriculture. In *IEEE International Conference on Wireless Communications, Networking and Information Security*, Beijing, China, pp. 571–575.

Huang, W., Guan, Q., Luo, J., Zhang, J., Zhao, J., Liang, D., ... Zhang, D. (2014). New optimized spectral indices for identifying and monitoring winter wheat diseases. *IEEE Journal of Selected Topics in Applied Earth Observations and Remote Sensing, 7*(6), 2516–2524.

Jayaraman, P. P., Palmer, D., Zaslavsky, A. & Georgakopoulos, D. (2015). Do-it-yourself digital agriculture applications with semantically enhanced IoT platform. In *IEEE Tenth International Conference on Intelligent Sensors, Sensor Networks and Information Processing (ISSNIP)*, Singapore, pp. 1–6.

Kalaivani, T., Allirani, A. & Priya, P. (2011). A survey on Zigbee based wireless sensor networks in agriculture. In *3rd International Conference on Trendz in Information Sciences & Computing (TISC'11)*, Chennai, pp. 85–89.

Khattab, A., Abdelgawad, A. & Yelmarthi, K. (2016). Design and implementation of a cloud-based IoT scheme for precision agriculture. In *28th International Conference on Microelectronics (ICM)*, Giza, pp. 201–204.

Konstantinos, K., Apostolos, X., Panagiotis, K. & George, S. (2007). Topology optimization in wireless sensor networks for precision agriculture applications. In *International Conference on Sensor Technologies and Applications*, Valencia, pp. 526–530.

Lee, M., Hwang, J. & Yoe, H. (2013). Agricultural production system based on IoT. In *16th International Conference on Computational Science and Engineering*, pp. 833–837. IEEE Publications, Sydney, NSW, Australia, NSW.

Li, S., Cui, J. & Li, Z. (2011). Wireless sensor network for precise agriculture monitoring. In *4th International Conference on Intelligent Computation Technology and Automation*, Shenzhen, Guangdong, pp. 307–310.

Li, Y., Guo, X., Shi, R. & Yang, F. (2014). Monitor and control wireless sensor nodes by B/S architecture, In *International Conference on Wireless Communication and Sensor Network*, pp. 204–206.

Liu, Y. et al. (2014). Enterprise-oriented iot name service for agriculture product supply chain management, In *International Conference on Identification, Information and Knowledge in the Internet of Things*, Beijing, China, pp. 237–241.

Markovic, D. et al. (2015). Application of IoT in monitoring and controlling agricultural production. *Actaagriculturaeserbica, 1531*(40), 145,

Mat, I., MohdKassim, M. R., Harun, A. N. & Mat Yusoff, I. (2016). IoT in precision agriculture applications using wireless moisture sensor network. In *IEEE Conference on Open Systems (ICOS)*, Langkawi, pp. 24–29.

Mewes, T., Franke, J. & Menz, G. (2009). Data reduction of hyper spectral remote sensing data for crop stress detection using different band selection methods. In *IEEE International Geosciences and Remote Sensing Symposium*, Cape Town, pp. III-463–III-466.

Mohanraj, I., Ashokumar, K. & Naren, J. (2016). Field monitoring and automation using IOT in agriculture domain. *Procedia Computer Science, 93,* 931–939. ISSN: 1877-0509.

MohdKassim, M. R., Mat, I. & Harun, A. N. (2014). Wireless Sensor Network in precision agriculture application. In *International Conference on Computer, Information and Telecommunication Systems (CITS)*, Jeju, pp. 1–5.

Patil, K. A. & Kale, N. R. (2016). A model for smart agriculture using IoT. In *International Conference on Global Trends in Signal Processing, Information Computing and Communication (ICGTSPICC)*, Jalgaon, India, pp. 543–545.

Qiu, T., Xiao, H. & Zhou, P. (2013). Framework and case studies of intelligence monitoring platform in facility agriculture ecosystem. In *2nd International Conference on Agro-Geoinformatics*, Fairfax, VA, pp. 522–525.

Ramundo, L., Taisch, M. & Terzi, S. (2016). State of the art of technology in the food sector value chain towards the IoT. In *IEEE 2nd International Forum on Research and Technologies for Society and Industry Leveraging a Better Tomorrow (RTSI)*, Bologna, pp. 1–6.

Ryu, M., Yun, J., Miao, T., Ahn, I. Y., Choi, S. C. & Kim, J. (2015). Design and implementation of a connected farm for smart farming system. *IEEE Sensors, Busan*, 1–4.

Shi, Y., Wang, Z., Wang, X. & Zhang, S. (2015). Internet of Things application to monitoring plant disease and insect pests. In *International Conference on Applied Science and Engineering Innovation*. Atlantis Press.

Sreekantha, D. K. & Kavya, A. M. (2017). Agricultural crop monitoring using IoT - A study. In *11th International Conference on Intelligent Systems and Control (ISCO)*, Coimbatore, pp. 134–139.

Tuli, A., Hasteer, N., Sharma, M. & Bansal, A. (2014). Framework to leverage cloud for the modernization of the Indian agriculture system. In *IEEE International Conference on Electro/Information Technology*, Milwaukee, WI, pp. 109–115.

Vatari, S., Bakshi, A. & Thakur, T. (2016). Green house by using IOT and cloud computing. In *IEEE International Conference on Recent Trends in Electronics, Information & Communication Technology (RTEICT)*, Bangalore, pp. 246–250.

Weimin Qiu, L. D., Wang, F. & Yan, H. (2014). Design of intelligent greenhouse environment monitoring system based on ZigBee and embedded technology In *IEEE International Conference on Consumer Electronics*, China, Shenzhen, pp. 1–3.

Yan, D.-e. (2011). Design of intelligent agriculture management information system based on IoT. In *4th International Conference on Intelligent Computation Technology and Automation*, Shenzhen, Guangdong, pp. 1045–1049.

Ye, J., Chen, B., Liu, Q. & Fang, Y. (2013). A precision agriculture management system based on Internet of Things and WebGIS. In *21st International Conference on Geoinformatics*, Kaifeng, pp. 1–5.

Yunping, C., Xiu, W. & Chunjiang, Z. (2009). A soil sampling intelligent system based on elastic algorithm and GIS. In *5th International Conference on Natural Computation*, Tianjin, pp. 202–206.

Zhang, J., Li, A., Li, J., Yang, Q. & Gang, C. (2011). Research of real-time image acquisition system based on ARM 7 for agricultural environmental monitoring. In *International Conference on Remote Sensing, Environment and Transportation Engineering*, Nanjing, China, pp. 6216–6220.

Zhang, M. & Qin, Z. (2004). Spectral analysis of tomato late blight infections for remote sensing of tomato disease stress in California. *IEEE International Geoscience and Remote Sensing Symposium, Anchorage, 6*, 4091–4094.

Zhao, J.c., Zhang, J.f., Feng, Y. & Guo, J.x. (2010). The study and application of the IOT technology in agriculture. In *3rd International Conference on Computer Science and Information Technology*, Chengdu, China, pp. 462–465.

7 Facial Feature-Based Human Emotion Detection Using Machine Learning
An Overview

Mritunjay Rai, Agha Asim Husain, Rohit Sharma, Tanmoy Maity and R. K. Yadav

CONTENTS

DOI: 10.1201/9781003097518-7

INTRODUCTION

Facial expression is one of the effective ways through which people can show their feelings. For human beings to communicate their emotions and intentions, the facial expression recognition system becomes one of the effective means. Facial recognition and detection are a technology that can recognise people based on their faces. It records, analyses and compares samples based on the detail of a human's face. The job of facial detection is an important step since it can identify and locate human faces in frames and videos. The face-capturing process is a conversion of analogue information which is the face as input into a set of digital information which is data based on the human facial features. With day-to-day advancement in technology, artificial intelligence (AI) has become extremely important in the identification and detection of human facial expressions. According to Market Global Forecast for the year 2025, the growth rate of AI will increase to US$190.61 billion compared to US$18.3 billion in 2019. That is why we can see its applications everywhere such as in phones or devices for security, home locks, to combat shoplifting, airports, etc.

In some cases, such as hospitalised patients or those with psychological conditions, people may be restricted from showing their emotions; therefore, a better understanding of human emotions will assist in constructive communication. Human emotion recognition has received widespread attention with the introduction of the Internet of Things (IoT) and smart technologies in hospitals, homes and other public places. Intelligent personal assistants (IPA) [1] such as Apple–Siri, Amazon–Alexa, Microsoft–Cortana, and Google–Assistant use natural language processing (NLP) to communicate with humans, but when emotionally excited, it improves the level of productive communication and human intelligence. Facial expression analysis is an impressive and challenging issue that will affect important applications in many fields, such as human computer interaction (HCI) and medical applications, etc. It is not only used to identify faces, but many companies are using facial recognition for security purposes. Immigration authorities in many countries use facial recognition to enforce border control, etc. The basic aspect of thermal cameras is the detection of temperature changes in the energy components of the electromagnetic spectrum. Non-invasive approaches (e.g., thermal cameras) are used to collect data. The thermal imaging (TI) system generates a thermal image of the body temperature profile of the surface. The temperature of the skin is correlated with the circulation of the blood and may be associated with physiological measures such as emotions. Planck's law is used to calculate body surface temperature, but the emissivity value of the skin is difficult to measure precisely [2].

Previous research often appears to believe that the emissivity of the skin surface is approximately 0.96 ± 0.003. During a human body temperature analysis, the body must be still, and the subject should not take any drugs, such as alcohol, and should not smoke before the test. These actions may cause temperature variations leading to incorrect calculations. Here are few cases listed where facial recognition plays an important role.

MAKING CAR PASSENGERS SAFE

Automobile manufacturers around the world are adding many features to improve driver safety. Car manufacturers are using AI to understand human emotions through facial features. Using facial emotion detection, smart cars can alert the driver when they are drowsy. This feature will avoid mishaps and add security and safety to the driver and passengers.

Roughly around 95% of lethal road accidents are related to driving mistakes as claimed by the US Department of Transportation. Facial identification can recognise changes in outward appearances due to fatigue and send a customised caution to the driver requesting that the driver stop for an espresso, change the music, change the temperature or take some fresh air to become refreshed for driving again.

FACIAL EXPRESSION DETECTION IN INTERVIEWS

The interaction between the candidate and the interviewer is susceptible to many categories of judgement and subjectivity. This subjectivity makes it difficult to determine whether the candidate's personality is suitable for the job. There are multiple levels of language interpretation, cognitive bias and context, so it is difficult to determine whatever the person is trying to express. And that is where AI plays a pivotal role, this can evaluate candidates' emotional states to gather their emotional responses and then further assess their physical features.

Now, it is possible to determine whether the candidate is replying to all questions honestly by measuring the candidate's emotional changes during the answering process and corresponding to a large amount of knowledge in the domain. Human resource tools can not only assist in formulating recruitment strategies but also help to design human resource policies that enable employees to achieve the best performance.

CORRECT PASSENGERS ARE PICKED UP BY TAXI DRIVERS

Passengers in the carpooling industry can also benefit from facial recognition. In the context of the sharing economy, facial recognition adds a layer of verification and security to passengers [3]. Drivers are also required to scan their faces on mobile devices to verify their identity and corresponding driver and security credentials. This information is delivered to their passengers so that they can be assured.

In the future, cars may be equipped with facial scanning technology, which can recognise human faces and unlock the vehicle to allow entry only to that passenger who originally booked the ride. This will bring additional safety advantages without the need for additional equipment.

VIDEO GAMES TESTING

Keeping the specific target audience in mind when designing any video games, every video game aims to induce a specific behaviour and a set of emotions of the user. In the testing stage, players are approached to play the game for a given time and give their feedback to determine the eventual outcome. Utilising facial feature location, the user is assisted to better understand the feelings that players experience while playing, without the need to physically dissect the total video.

UNLOCKING PHONES

Facial recognition is presently being used to unlock a variety of mobile phones. This computerised technique is an effective way to protect personal information and to ensure that criminals cannot access vital information even if the phone is stolen.

INVESTIGATION

Facial recognition can also aid investigations by identifying faces in surveillance images or recordings. Several investigating agencies use facial recognition software to identify people at crime scenes.

CLASS ATTENDANCE

Facial recognition helps track the attendance of students. Nowadays many schools, colleges and universities use tablets or any other face-scanning devices to scan students' faces and match their photographs with databases to verify their identities.

AIRPORT SECURITY

All airports are nowadays safer due to the use of facial recognition for the passengers, staff and aircraft crew members. Airbus staff have started using facial recognition to help check people's identity, luggage, etc. This helps passengers to board the plane quickly.

HOTELS, CASINOS AND BARS

Facial recognition can help hotels, casinos and bars to identify defaulters from entering their premises. Also, facial recognition can identify members on the voluntary exclusion list. It would be easy for the owners to call nearby police to handle the defaulters or non-payers.

Adding many other examples here is beyond the scope of this chapter. The remaining structure of the chapter is organised into three sections, covering the literature review, the best possible methodologies for the detection of human emotions and a detailed discussion on the work done so far, as well as prospects for the future.

LITERATURE REVIEW

The recognition and analysis of human emotions have a wide range of significance in various domains of security, medical, etc. application. Human–Machine Interaction

(HMI), intelligent tutoring systems (ITS), content-based frame and video retrieval, acute surroundings and vehicle driver monitoring inspection [4]. Psychiatrists and psychologists determine various human mental health conditions by using emotion recognition systems. Human emotions can be categorised by the process of recognising emotions through facial impressions, speech, various physiological parameters, body gestures and movements. In the thermal infrared spectrum, emotion recognition using facial expression has received less attention than that for the visible spectrum. For the past few decades, enough methods and algorithms have been suggested by various scientists and researchers to distinguish emotions predicted from the expressions of human faces and the audio signals. Due to the complexity in prediction, even with the involvement of AI, computer vision, psychological and physiological parameters, it remains a challenging task. According to the existing literature by numerous scientists and researchers, it is believed that human emotion recognition through facial expressions is the most influential method. However, while using facial expression features it is difficult to perceive human emotions because of external noises, such as varying light conditions and complex head motion [5]. Also, the results obtained need to be improved for further emotion classification.

This chapter addresses the primary six classes of facial emotional conditions, i.e., joy, sorrow, anger, disgust, shock and unemotional. Khan et al. [6] suggested that by using multiple variable tests and linear discriminant analysis (LDA) the varying thermal intensity values of facial features [7] helps in categorising the intentional facial expression. Other research by Trujillo et al. [8] suggested confining facial attributes or features in thermal frames through an approach to interest point identification and clustering. Hernández et al. [9] presented the gray-level co-occurrence matrix (GLCM) as a zonal descriptor for infrared pictures and described an approach that picks the region of interest (ROI) and GLCM variables simultaneously using an initial computation technique. Jain et al. [10] suggested a device based on thermal imaging (TI) technology for non-intrusive lie detection. To categorise ROI and extract attributes for implicit information inference, face tracking is directly used on thermal videos or images. Liu and Wang [11] suggested a technique of emotion identification and recognition from the facial temporal frame sequence. The five statistical characteristics and features of the temperature variation histogram are derived from the matrix of the facial temperature variation. Also, as a classifier, the distinct hidden Markov models (HMM) are used. Eom and Sohn [12] analysed variations in facial temperature caused by facial expression and emotional condition to identify the emotion of an individual using facial thermal videos or images. Another researcher named Bernhard Anzengruber [13] investigated the possible uses of TI systems to understand the mental states of drivers unobtrusively. This information is also used to provide immediate response during high-risk conditions. Krzywicki et al. [14] have suggested a direct correlation between thermal radiation in the human face and the person's physiological stress levels.

Merla and Romani [15] selected three basic emotional states: stress, fear and the stimulation of pleasure, and respectively examined the thermal signatures of the faces of many healthy participants. Pavlidis and Levine [16] are the first to propose a thermal image-based human emotion recognition system. It uses heat transfer modelling and thermal data is converted to the blood circulation rate. The subject is then classified as deceptive or non-deceptive using the nearest-neighbour classification system. The temperature information from thermal images is used for human emotion recognition. All this research suggests that infrared images provide useful information for the detection of facial expressions and helps in predicting the state of emotion. The feature used in early research consists of the statistical parameters of the temperature of the ROI taken by the infrared camera, or other imaging functions designed to reflect the image of the visible spectrum instead of the thermal infrared spectrum. Frames captured in the visible spectrum present the appearance and geometric details of a face; in contrast, the thermal images show the facial degree of temperature distribution. Therefore, the features developed for visible frames cannot completely capture the degree of temperature distribution contained in thermal images. The characteristics of thermal images are somehow shown by temperature statistical parameters. Due to this, it becomes difficult to obtain implicit data like geometric and appearance from the degree of temperature distribution. So far, there are no manual functions built for thermal images.

The existing work related to human emotion detection describes the usefulness of deep learning in a variety of areas, like machine vision, speech identification and recognition, etc. However, there has been limited work done so far in the domain of facial impression or expression and emotional identification and recognition. Ranzato et al. [17] used a Markov random field (MRF) as the lowest layer of a deep belief network (DBN) for the analysis of facial expression. Many researchers have proposed different theories to understand the different facial expressions and can handle the occluded areas by filling occlusions. DBN was introduced by Susskind et al. [18] to learn and recognise facial expressions. Through unsupervised learning, many face images with only scattered labels are fully utilised. To solve the challenge of emotion detection, Rifai et al. [19] proposed a method based on the multi-scale contractive convolutional network (CCNET) and contractive discriminative analysis (CDA). These feature-extracting networks are used to filter the variation induced by facial posture and facial identity or by morphology. Sabzevari et al. [20] have suggested a technique to extract the facial expressions of the faces through a combined approach of facial maps and DBN. Above all, studies have focused on visible frames. Kim et al. [21] studied the utility of deep-learning technology for the recognition of emotions based on facial features. This is an unsupervised learning technique used in the audio-visual emotion recognition system. For speech emotion recognition, Stuhlsatz et al. [22] have suggested a unified discriminative approach based on deep neural networks (DNN). Schmidt et al. [23] investigated feature learning using the aggregated amplitude spectrum and multiple timescales. To learn features directly from the amplitude spectrum, they used a regression-based DBN. Martinez et al. [24] have developed an application based on deep-learning approaches to represent the physiological effects.

There is a huge amount of evidence demonstrating the use of infrared imaging technology in the healthcare sector. For example, infant behavioural responses in happy and peaceful emotions, emotional response in social interactions, identification of skin temperature, behavioural patterns after lower limb training and assessment of a young person's stress.

An ample amount of research has been done in the field of TI technology to understand human sensations and emotional states. There are emotions in everyday life. For instance, if medical students are not prepared to face challenging emotional nursing tasks, they may feel vulnerable. The use of TI technology is the latest advancement in the measurement of the emotional state. TI technology can easily measure the reflected radiation that can be used to calculate the temperature. Researchers working on human emotion detection are interested in using TI techniques as they can accurately measure the temperature of the skin surface without the use of cords or electrodes. Investigation on infrared TI techniques used to measure emotions comes from several disciplines. Psychotherapists have documented a correlation between facial temperature changes and arousal in humans and non-human primates.

METHODOLOGY

For the development of society and industry, there is a drastic increase in the demand for smart technologies. Also, there is a tremendous increase in technologies that can assess the needs of prospective clients and provide the best possible solutions for them. The automated assessment of human emotion is important in some of the application areas like robotics and artificial intelligence, advertising, arts, and crafts and the entertainment industry, etc. These applications are used to achieve various targets:

- To develop intelligent collaborative or smart machines which can interact with humans, for automated industries.
- To create customer-based specialised advertisements that depend on the emotional state, for marketing firms.
- To improve the learning processes, perception methodologies and knowledge transfer in the education industry.
- To suggest another appropriate multimedia service for the intended viewers.

The various scientific literature presents different methods for the classification of human emotions and defines the boundaries between mood, affect and emotions. Depending on the classification, the primary terms related to human emotions are defined as below:

- The term 'emotion' refers to a living organism's reaction to a certain stimulus against any situation or incident. Normally, this is a strong, brief sensation that the person is aware of.
- The 'affect' is the result of the influence of feelings and sentiments, and it involves the complex relation of emotion.
- The 'feeling' is often interpreted for a particular subject that the individual is conscious of; its longevity determines the time duration that the image of the object remains in the person's memory.
- The 'mood' appears to become more subtle, long-lasting, slight-severe, extra in the background, but that affects an individual's emotional state in either a positive or a negative way.

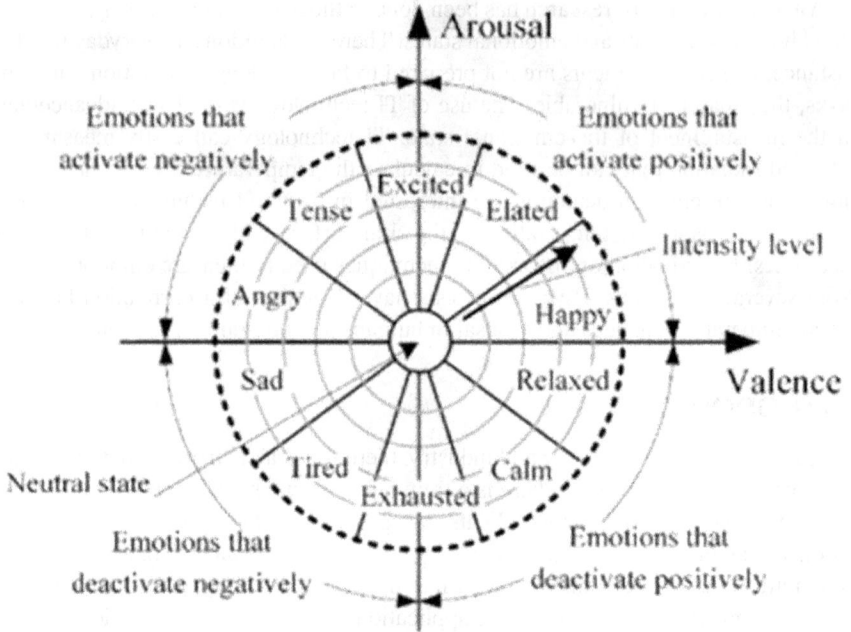

FIGURE 7.1   Russell's circumplex prototype of emotions [24].

Most of the researchers employ Russel's circumplex prototype of emotions, as shown in Figure 7.1 [24], which provides a two-dimensional projection of basic emotions for valence and anxiety. Using this method, it is possible to detect the desired emotion and its intensity by evaluating only two dimensions.

The classification and evaluation of emotions have become simple using the above-mentioned model. However, numerous challenges remain with the evaluation of emotions, the choice of methods for monitoring and assessing output, and the choice of testing hardware and software. Furthermore, the multidisciplinary approach to the problem of emotion identification and assessment makes it difficult to solve. Cognitive research focuses on emotion recognition and intensity assessment whereas examination and analysis of human body variables are correlated to medical science and engineering. Intelligent machines with human emotions are likely to make the community a better place. AI technology, particularly machine learning, along with machine vision, is making progress in understanding human facial emotions (see Figure 7.2).

The following are the three main components of facial emotion detection:

i. Image reprocessing.
ii. Feature extraction.
iii. Feature classification.

The proposed approach is described by a combined analysis of spatial information that efficiently represents emotional states. Face recognition and emotion

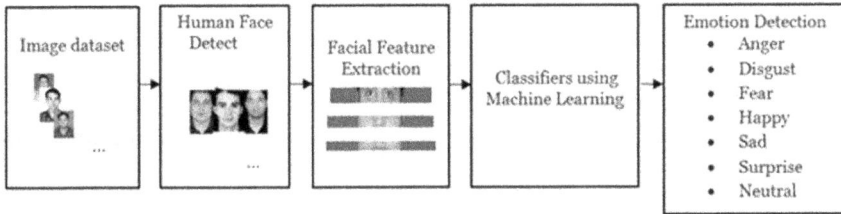

FIGURE 7.2   Facial expression system.

classification features are combined into a single system. The model describes the facial image's spatial patterns while considering the direct association between adjacent areas. Face detection, visual representation and expression identification are the three elements that define the structure. As a result, any of the modern state algorithms can be used to replace any specific component. Offline training is possible for individual components. Face recognition techniques are classified into three groups based on their intensity:

a) Feature-based approach.
b) Holistic-based approach.
c) Hybrid approaches.

A) FEATURE-BASED APPROACH

This approach is further divided into two major approaches based on key points and local appearances. The feature-based approach uses the image as input to recognise and remove distinguishing facial characteristics including the eyes, mouth and nose. Then, to reduce the input facial image to a vector of geometric features, computes the spatial relationship between those facial points [25]. Scale-invariant feature transform (SIFT), speeded-up robust features (SURF), binary robust independent elementary feature (BRIEF), local binary pattern (LBP), histogram of oriented gradients (HOG), elastic bunch graph matching (EBGM), local phase quantisation (LPQ) and others are methods that contain many design approaches. The following are the two methods that will be discussed:

(i) Geometric feature-matching technology

It is described as the process of extracting a set of facial traits from a photograph. A vector defines the layout settings. This vector depicts the nose, eyes, brows, mouth, chin and overall contour of the face in terms of size and placement. This strategy's strength is that it aids in overcoming the oversight issue. The disadvantages of this strategy include its computational complexity and lack of stability.

(ii) Elastic bunch graph

Elastic bunch graph matching (EBGM) is capable of categorising objects with general structures, such as faces with the same pose. This design is composed of a complex relationship structure. It defines a set of reference points on the face to generate a single face graph, each point of reference being a node on the whole face. To mark the specific graphics, the appropriate response of the Gabor filter is used. Each arch should be marked with the difference between the corresponding reference points [26]. A bunch graph is formed by combining the whole sets of these graphs into a layered structure. The image map of a new facial picture is compared to the image maps of all existing facial images and the highest image to identify it. The significance of the similarity shall be determined as the closed match [27]. Feature points are found in feature-based techniques before analysis to align images with images of known individuals. The technology focused on features that can be kept constant in size, orientation and emission. It has the compactness of facial image expression and high-speed image matching. Feature-based technology requires the ability to recognise itself as it is hard to auto-detect features.

B) HOLISTIC-BASED APPROACH

The methodology is further divided into analytical and computational modelling approaches. These can be linear and non-linear techniques. The holistic-based approach mainly uses image representations for the detection and recognition of facial features. The observation is based on a detailed analysis of the entire frame instead of facial features. The holistic-based approaches consist of some methods like principal component analysis (PCA), LDA, kernel-PCA, DNN, support vector machine (SVM) and Eigen face (EF).

(i) Statistical approach

This is the basic technique used for the detection and extraction of facial features. During the classification of the emotional state, the frame sequence from video data is interpreted using a binary sequence of pixel intensity values. The detection is performed by analysing the facial features of the input face image (testing image) and the trained facial images from the dataset. Some of the statistical approaches for facial feature extraction are LDA, Pearson's correlation coefficient (PCC), Kullback, Fisher linear discriminant (FLD) and Eigen surface [28], etc. Among all the Eigen surface-based approaches is one of the most widely used facial recognition techniques. This technique uses PCA as the main component to reduce the dimension and is also known as the Karhunen–Loeve (KL) transform. The PCA is used for the identification and detection of the face [29]. The PCA-based model has ease of implementation and is highly effective due to the short execution time in image dimension reduction. The PCA also maintains a high correlation among both training data and recognition data [30]. Accuracy deteriorated a lot due to various factors, including lighting, which is one such drawback.

PCA is one of the best-known methods used in facial detection and classification. Data is analysed in this technique in a multivariate statistical analysis pattern. PCA transforms the original image to a presentable form in which each attribute is linearly independent via a linear transformation. PCA is frequently used to describe the key components of data and reduces the dimensionality of high-dimensional data. The main advantages of PCA are to remove the correlated features of the dataset, improve the efficiency of the algorithm, reduce the issue of overfitting and improve the visualisation of the dataset.

(ii) Artificial intelligence methods

Artificial intelligence (AI)-based methodologies primarily use the concept of machine-learning (ML) and deep neural network (DNN) technologies. These technologies play a vital role in the classification of facial features-based human emotions. The AI methods are not limited to random forest (RF), linear regression (LR), decision tree (DT), principal component analysis (PCA) and support vector machine (SVM). The SVM is part of supervised ML algorithms and is an efficient method for LR and classification of data. The main motive of SVM is to find the hyperplane that best divides the two categories. Based on this hyperplane, it is optimal to classify the new data into the class to which the new data belongs.

This technique has the advantage that it does not damage any data, as it focuses only on selected regions in the image. However, this technology is relatively costly in terms of computation and often involves a strong correlation between the test images and the training images.

c) Hybrid approach

The hybrid approach is a fusion of both a feature and a holistic design to accomplish better outcomes. The two approaches are being used in a way that the limitation of one can be compensated by the advantage of another approach, such that the overall accuracy of the results is improved. One such example is the combination of a template-matching algorithm and 2D PCA algorithm [32, 33].

Among the three categories discussed above for facial emotion identification, this chapter mainly focuses on the holistic-based approach. With the advancement in ML-based algorithms, emotion detection using facial expressions finds its real-time-based application in a wider domain. These domains include surveillance at borders, airports, public places, etc. This approach also helps in the development of intelligent traffic surveillance systems [34] and advanced intelligent video surveillance systems (AIVSS) [35].

DISCUSSION AND CONCLUSION

Emotion recognition becomes an immensely powerful and effective method for evaluating human thoughts and emotions. It forecasts human activities to provide more appropriate marketing content for the advertising industry as well as professional training firms. The recognition and assessment of emotional states are indeed highly effective in the evaluation of numerous HMI systems. The emotional states and human body responses are interdependent and have been well-defined for ages. Many challenges remain in the development of automated technology for monitoring and analysing the data. The incorporation of ML for data processing is an extraordinarily powerful combination. It might advance the practical implementation in all the application areas. The advancement not only starts with the marketing and advertising domains but also helps in the advancement of industrial engineering applications.

This chapter provides a brief discussion on human emotion detection which gives additional information and security about the person. Several methods have been discussed for the detection of human emotions along with the face. With the advancement in new technologies like ML and DL, emotion features are extracted, and it turns out that both algorithms are good at pattern recognition and classification.

REFERENCES

1. Lopatovska, I. (2019). Inspection of digital intellectual personal assistants. *Ukrainian Journal of Library Science and Information Sciences*, *3*(3), 72–79. doi: 10.31866/2616-7654.3.2019.169669
2. Rai, M., Maity, T. & Yadav, R. K. (2017). Thermal imaging system and its real-time applications: A survey. *Journal of Engineering Technology*, *6*(2), 290–303.
3. Hu, X. & Wang, J. (2018). Taxi Driver's operation behaviour and passengers' demand analysis based on GPS data. *Journal of Advanced Transportation*, *2018*, 11, Article ID 6197549. doi: 10.1155/2018/6197549
4. https://www.facefirst.com/blog/amazing-uses-for-face-recognition-facial-recognition-use-cases
5. https://towardsdatascience.com/facial-emotion-detection-using-ai-use-cases-248b932200d6
6. Oz, I. & Khan, M. M. (2012). Efficacy of biophysiological measurements at ftfps for facial expression classification: A validation. In *Proceedings of the 2012 IEEE-EMBS International Conference on Biomedical and Health Informatics*, pp. 108–111.
7. Rai, M. & Yadav, R. K. (2016). A novel method for detection and extraction of human face for video surveillance applications. *International Journal of Signal and Imaging Systems Engineering*, *9*(3), 165–173.
8. Trujillo, L., Olague, G., Hammoud, R. & Hernandez, B. (2005). Automatic feature localization in thermal images for facial expression recognition. In *Proceedings of the 2005 IEEE Conference on Computer Vision and Pattern Recognition*, p. 14.
9. Hernández, B., Olague, G., Hammoud, R., Trujillo, L. & Romero, E. (2007). Visual learning of texture descriptors for facial expression recognition in thermal imagery. In *Computer Vision and Image Understanding*.
10. Jain, U., Tan, B. & Li, Q. (2012). Concealed knowledge identification using facial thermal imaging. In *Proceedings of the 2012 IEEE International Conference on Acoustics, Speech and Signal Processing*, pp. 1677–1680.
11. Liu, Z. & Wang, S. (2011). Emotion recognition using hidden Markov models from facial temperature sequence. *Lecture Notes in Computer Science*, *6975*, 240–247
12. Eom, J. S. & Sohn, J. H. (2012). Emotion recognition using facial thermal images. *Journal of the Ergonomics Society of Korea*, *31*(3), 427–435
13. Bernhard Anzengruber, A. R. (2012). "Face light": Potentials and drawbacks of thermal imaging to infer driver stress. In *Proceedings of the 4th international conference on Automotive User Interfaces and Interactive Vehicular Applications*, pp. 209–216.
14. Krzywicki, A. T., He, G. & O'Kane, B. L. (2009). Analysis of facial thermal variations in response to emotion: Eliciting film clips. In *Proceedings of the SPIE Defense, Security, and Sensing*, *7343*, 73412-1–73411-11.

15. Merla, A. & Romani, G. L. (2007). Thermal signatures of emotional arousal: A functional infrared imaging study. In *Proceedings of the 29th Annual International Conference of the Engineering in Medicine and Biology Society*, pp. 247–249.

16. Pavlidis, I. & Levine, J. (2002). Thermal image analysis for polygraph testing. *IEEE Engineering in Medicine and Biology Magazine*, 21(6), 56–64.

17. Ranzato, M., Mnih, V., Susskind, J. & Hinton, G. (2013). Modeling natural images using gated MRFs. *IEEE Transactions on Pattern Analysis and Machine Intelligence*.

18. Susskind, Joshua M., Hinton, G. E., Movellan, J. R. & Anderson, (May 1st 2008). Generating facial expressions with deep belief nets. *Affective Computing*, Jimmy Or, IntechOpen, doi: 10.5772/6167. Available from: https://www.intechopen.com/chapters/5197

19. Rifai, S., Bengio, Y., Courville, A., Vincent, P. & Mirza, M. (2012). Disentangling factors of variation for facial expression recognition. In *Proceedings of the 12th European Conference on Computer Vision*, 6, pp. 808–822

20. Sabzevari, M., Toosizadeh, S., Quchani, S. & Abrishami, V. (2010). A fast and accurate facial expression synthesis system for color face images using face graph and deep belief network. In: *Proceedings of the 2010 International Conference on Electronics and Information Engineering*, V2, 354–358.

21. Kim, Y., Lee, H. & Provost, E. M. (2013). Deep learning for robust feature generation in audio-visual emotion recognition. In: *Proceedings of the 2013 IEEE International Conference on Acoustics, Speech and Signal Processing*

22. Stuhlsatz, A., Meyer, C., Eyben, F., Zielke, T., Meier, G. & Schuller, B. (2011). Deep neural networks for acoustic emotion recognition: raising the benchmarks. In: *Proceedings of the 2011 IEEE International Conference on acoustics, Speech and Signal Processing*, pp. 5688–5691.

23. Schmidt, E. M., Scott, J. J. & Kim, Y. E. (2012). Feature learning in dynamic environments: Modeling the acoustic structure of musical emotion. In: *Proceedings of the International Society for Music Information Retrieval*, pp. 325–330

24. Martinez, H., Bengio, Y. & Yannakakis, G. (2013). Learning deep physiological models of affect. *IEEE Computational Intelligence Magazine*, 8(2), 20–33.

25. Newman, B. M. & Newman, P. R. (2020). *Theories of Adolescent Development*. Academic Press, pp. 279–312, ISBN 9780128154502, https://doi.org/10.1016/B978-0-12-815450-2.00010-3.

26. Jafri, R. & Arabnia, H. R. (2009). A survey of face recognition techniques. *Journal of Information Processing Systems*, 5, 41–63.

27. Sharif, M., Mohsin, S. & Javed, M. Y. (2012). Face recognition techniques. *Research Journal of Applied Science Engineering and Technology*, 4, 4979–4990.

28. Masupha, L., Zuva, T., Ngwira, S. & Esan, O. (2015). Face recognition techniques, their advantages, disadvantages, and performance evaluation. In *2015 International Conference on Computing, Communication and Security (ICCCS)*. doi: 10.1109/cccs.2015.7374154

29. Draper, B. A., Baek, K., Bartlett, M. S. & Beveridge, J. R. (2003). Recognizing faces with PCA and ICA. *Computer Vision and Image Understanding*, 91, 115–137.

30. Vijayakumari, V. (2013). Face recognition techniques: A survey. *World Journal of Computer Application and Technology*, 1.

31. Mir, A. & Mir, A. G. (2013). Feature extraction methods (PCA fused with DCT). *International Journal of Advances in Engineering and Technology*, 6, 2145–2152.

32. Wang, J. & Yang, H. (2008). Face detection based on template matching and 2DPCA algorithm, presented at the congress on image and signal processing.

33. Agarwal, M., Jain, N., Kumar, M. & Agrawal, H. (2010). Face recognition using eigen faces and artificial neural network. *International Journal of Computer Theory and Engineering*, 2, 624–629.
34. Husain, A. A., Maity, T. & Yadav, R.K. (2020). Vehicle detection in intelligent transport system under a hazy environment: a survey. *IET Image Processing*, *14*(1), 1–10.
35. Rai, M., Husain, A. A., Maity, T., et al. (2018). *Advance Intelligent Video Surveillance System (AIVSS): A Future Aspect.* Video Surveillance Intech Open, London, UK.

8 A Comprehensive Study of Security in Electronic Commerce Frameworks in the Modern Era of Technology

Sushree Bibhuprada B. Priyadarshini

CONTENTS

DOI: 10.1201/9781003097518-8

BACKGROUND STUDY

Internet business or electronic trade could be a technique of ongoing exchange that serves the needs of merchants, clients and business associations, to decrease the value and upgrade the standard of items while raising the speed of conveyance. Online business conducted through the paperless trade of corporate data include 'electronic data exchange' (EDI), 'electronic mail' (email), 'electronic bulletin boards', 'electronic funds transfer' (EFT) and other network-based advances [1].

The great result of an electronic trade development relies upon endless segments, just as any not confined to the plan of action relating to groups like buyers, financial specialists, the product and the security of data transmissions and capacity. Data integrity has been recognised since a lot of attacks have disabled notable sites; these have occurred because of the impersonation of Microsoft staff, and furthermore, the misuse of charge cards of buyers at business-to-shopper e-commerce sites. Security is on the mind of every web-based business specialist who creates or transmits any information that will be sensitive if it is lost [1, 2]. An arms race is developing that is shaped by new security requirements forced by those who are working to test the security of the system.

Encryption is one of the main methods of making certain that data security and information is trustworthy. Encryption is a general word for a process that scrambles data rendering it unreadable and thus enabling the data to be safely transmitted through the web. Encryption will guarantee the data at the highest level by restricting individuals from accessing the data. This situation occurs when someone intrudes the corresponding data and controls to sell out any client ID associated with it [3].

Additionally, advances in encryption will similarly promoted, by the demonstration of establishing the presence of clients, dealing with the illicit transmission that confirms the integrity of the data and ensures that clients take responsibility for data that they should be sent. Further, encryption will be required to stop people being worried about integrity issue arising in case of data transmissions. Also, 'open key encryption' [1, 2], or deviated encryption, is more crucial for the ultimate goal of internet business than symmetric encryption [3].

The huge improvement triggered by public key encryption was the introduction of the second key – which uses required features in the states of defensive trustworthiness of data. Open key encryption relies upon two keys, one that is open and one that is private. You cannot use one key on the off-chance that you have got the other. It attempts to attain an open key and that key may be provided to anyone for whom data needs to be transmitted [1–3]. You have my open key and use it to figure out a message. You transmit that message, and access the system. Anyone who sees the message cannot read it; therefore, they need the public key. The message is readable once the person acquires it, as I even have the sole duplicate of the private key, that will do the decoding process.

The most recognisable use of 'open key encryption' for electronic trade includes the work of faulty 'computerised certificates' [1] communicated by trusted outsiders. It is the responsibility of the individual. Let us assume you are a customer of major bank and you wish to collaborate with your bank. When you conduct a transaction with the bank, you may feel uneasy that the information may get blocked in transit and that the bank may think it was not you who transmitted the information. You and the bank could work with a trusted third party to help you encrypt the data over the web [2, 3].

The most unmistakable abuse of 'open key encryption' for electronic exchange involves defective 'automated certificates' [4] created by hackers. It is a matter of individual responsibility. Imagine that you are a client of a major bank and you wish to work with your bank. If conduct a transaction with the bank, you may worry that the data may get intercepted in transit and you may be concerned that the bank would not believe and it was you who transmitted the data. You and the bank collaborate while utilising a trusted third party to assist to transact over the web.

In any case, the key inside the transaction will decode the message that you transmitted so that everyone in the bank needs their own key to check the message. Those keys act instantly and the bank will be assured of two things: that you were the person who transmitted the message and that the only other party to the message was the bank. The funds were moved, as mentioned in [3] for data transmittal.

Public key encryption works within a closed system in which it does not make a difference if the system (n/w) is unstable. Despite the fact that on account of a distributed n/w like the web and the data go through a few hands, in the method of switches or switches and centres and data may be caught; the encryption keeps the data scrambled, except if the other party has the private key. Public key infrastructure (PKI) accepts the possibility that the least complex approaches to set up an arrangement of secure exchanges over systems are to decide a foundation which will strengthen open key encryption.

The PKI would create a situation at any place with any web client that may assign certificates that set up a secure transaction. The validation of keys may turn out to be easy and straightforward. Some web-based business users suggest that development of a smooth and solid PKI would have a significant role in boosting the development of internet business [4].

SIGNIFICANCE OF ELECTRONIC COMMERCE

The following are the major features of e-commerce:

- **Cashless payments:** internet business enables the use of all types of cards, electronic transfers through online banking and other electronic payment methods [1].

- **Service availability:** e-commerce focuses on automating the businesses of trades and services that they provide services to purchasers and be available anytime, anywhere. That means it offers twenty-four hours of services each and every day.
- **Advertising/marketing:** e-commerce helps in growing the advertising market for corporate goods and services. It helps to promote goods and services.
- **Improved sales:** requests for the products can be created, at any time and in any place without human mediation, utilising e-commerce. Ability to search for an item that has to be scaled up while enhancing its pattern and dealings is considered [1].
- **Support:** web-based business provides a range of approaches to give pre-purchase and post-purchase support, and in offering better customer service [1].
- **Inventory management:** the management of drug inventories is incredibly economical and easy to take care of.
- **Communication improvement:** e-commerce offers ways to connect more easily, securely and effectively with consumers and partners [1].

A COMPARATIVE STUDY OF TRADITIONAL COMMERCE VERSUS E-COMMERCE

Sl. No.	Traditional Commerce	E-Commerce
i.	Heavy reliance on the exchange of data between individual and entity.	The sharing of knowledge is generated directly across communication networks, having very little reliance on the exchange of personal data.
ii.	Communication includes a synchronous approach for sending data. For any manual communication, intervention is needed.	Communication normally occurs in synchronous mode. The electronics systems operate such that they can handle the communication appropriately.
iii.	Determining and preserving commonplace activities in conventional trade is difficult.	In e-commerce, a very uniform strategy is defined simultaneously and maintained.
iv.	Communication between businesses depends on personal skills.	There is no human intervention in e-commerce or the electronic economy.
v.	The unavailability of a forum in conventional trade, it relies on personal contact.	E-commerce provides users with a website where all information can be accessed in one place.
vi.	There is no standard information-sharing network since it is highly reliant on personal contact.	E-commerce facilitates a digital platform to assist in regional company operations.

PREREQUISITE TO E-COMMERCE SECURITY

As you engage in business on the web, you are going to experience three particular kinds of individual: (a) individuals who wish to search you, (b) individuals who wish to take from you, (c) individuals who wish to take from individuals who receive from you. The logical inconsistency that each site proprietor faces is that you, basically, wish to welcome the various individuals wholeheartedly. In a traditional store, it is easy to detect any trouble [3]. Working together online, nonetheless, proposes that you lose that essential intuition [3]. Security ought to be your priority when you are transacting on the web through your own site. It is, furthermore, a situation whereby you essentially cannot organise your on-line security by experimentation. You must get it right the first time, by virtue of learning from mistakes, which is hard for large businesses, and pretty much impractical for an independent venture. On the off-chance that you experience a security breach and it is analysed, you will need to recover your business and your reputation from scratch [5].

Many pages have confusing structures to be navigated before a customer will make even their first purchase. For the most part, these sites contain a wide range of information that is not relevant to the purchase. This is often in some cases the fault of the head office which is attempting to get CRM data and the issue is that the presale is not the correct time to collect that kind of data.

Legally you have an obligation to protect the data that you store that relates to your clients. There are even certain categories of data that you are not legally permitted to store. All things considered, a few sites do store that data that they are not permitted to store. It is better for you not to attempt this. In the pre-contract process, there will be the possibility of losing clients by mentioning an inordinate amount of information. Customers will go to another site where purchasing is less complex and any place where they will not be confronted a barrage of questions. Individuals are sharing a lot of data online, so your objective should be to gather the minimum amount of information necessary to generate trust [3, 4].

Preparing exchanges in-house will save a little money on each sale. If you make low volume, high cost sales, investment funds might be essential. There are a lot of advantages for misuse of external provider networks like PayPal, Skill and Worldpay. The preferred position is that you are not accessing financial information from your customer and importantly, not accumulating any information. That downside will affect elective things upon which ecommerce relies.

Facebook, Twitter, Yahoo, Gmail and numerous other organisations will cause problems for you once you travel outside your home country and do not have worldwide roaming activated on your telephone [4]. From remote locations, signing in from various gadgets can cause you problems, but in any event none of those organisations have direct power over your income.

Service providers provide the genuine outcomes at the time of need. Clients are notorious for not completing instructions appropriately; when you are unable send their items because of this, they will accuse you and that may lead issues like refunds, and after some time this could have an impact [1–5] on your business, and, moreover, your reputation. Assigning various information to specialised third party, might be marginally unpredictable for certain organisations. if you sell advanced items like e-books to clients who typically expect to receive their items in a split second. If you are showcasing physical products for sale, you have to set aside some additional effort to find out everything, and you should utilise it.

Check that the amounts, costs and item descriptions are correct. Additionally, check that any rebate or coupon codes are legitimate. As you will have the option to see, remaining secure does not require a great deal of effort or cost. It basically leads to ignoring the propensities for enormous firms in case of application aspect. In short: do not spy on your customers, do not collect information you do not need, secure the information what you have gathered, if feasible, by using a third-party provider and review it before you ship the items. Another issue you should consistently monitor is to ensure discounts correspond with the original offer. It has been known for individuals to get a purchase at a reduced price and a discount on the full value, and representatives do not generally notice [3–5].

Security Systems in E-Commerce

Security is a critical part of any internet transaction. Customers could lose trust in e-business if their security is compromised [1].

The essential needs for secure e-transactions are, accordingly:

- **Confidentiality:** it would not be beneficial for the organisation if unauthorised individuals were to access data. A transaction must not be disrupted.
- **Integrity:** data should not be modified over the network during transmission.
- **Availability:** data should be accessible whenever it is needed within a defined time period.
- **Authenticity:** there should be a mechanism for certifying a user prior to giving them access to the defined data.
- **Non-reputability:** it is the protection from being denied requests. At the point when a sender communicates something specific, the sender must not be set up so the message can be refused. Similarly, the message recipient must not be prepared to refuse the receipt.
- **Encryption:** data can only be encrypted and decrypted by an approved person.
- **Auditability:** in order for the data to be audited for integrity criteria, data should be registered.

SECURITY MEASURES IN E-COMMERCE

The main safety measures are as listed below:

- **Encryption:** this represents a very persuasive as well as effective way to protect data transmitted across the network. The sender of data encrypts the data with a code and only the intended receiver may decrypt the details with the same or a secret code.
- **Digital signature:** it ensures authenticity of the data. An e-signature authenticated with encryption and a password is a digital signature.
- **Security certificate:** it may be a distinctive digital ID programmed to check a personal website or a user's identity.

ROLE OF E-COMMERCE IN DAILY LIFE

There are various roles in e-commerce in daily life which operate in conjunction with the foundations started within the spheres of trade facilitation, cross-border transfer services, cybersecurity, daily customer interaction, international internet connectivity, etc. [1–7].

ROLE OF E-COMMERCE IN TRADE FACILITATION

This requires a number of duties to ensure that trade-related activities taking place online are carried out promptly and reasonably, as well as in relation to the outcome:

- Electronic signatures, which are in electronic form, are not to be denied solely on the basis of the signature.
- Authentication of electronic transactions, and the opportunity to consent to legal requirements.
- The legal basis of a party for electronic transactions, which should be compatible with the specific principles of e-commerce.
- Encourage online documentation and electronic submission of import and export documents.

ROLE OF E-COMMERCE IN CROSS-BORDER TRANSFER SERVICES

The capacity to move data across borders is significant to businesses that rely on online reservation networks, and telecommunications companies looking to send data seamlessly to the management of organisations across the markets of the 'Trans-Pacific Partnership' (TPP) [4]. For the first transaction in an exchange transaction, the TPP members resolved to allow a 'secured individual' (service providers and financial specialists) to move data across borders by electronic means. The accessibility requires that each jurisdiction can have its own regulatory necessities for the exchange of information. Governments could force conditions or limitations on the cross-border transfer of information that is desired to attain open-arrangement destinations, provided those measures are not applied in a way that will promote self-interest or separation or a hidden limitation on trade [2].

ROLE OF E-COMMERCE IN CYBERSECURITY

Shoppers and businesses need a secure and well-functioning internet to create the most important opportunities online. The TPP parties recognise the importance of cybersecurity cooperation through the national 'computer emergency response team' (CERT) network [4].

ROLE OF E-COMMERCE IN DAILY CUSTOMER INTERACTION

E-commerce recognises the advantages for customers' open access to the web within the framework of the policies, laws and regulations of every TPP party. These incorporate the ability to access internet services and applications, connect end-user devices to the internet and access data related to their internet access service providers' network management practices [2].

ROLE OF E-COMMERCE IN INTERNATIONAL INTERNET CONNECTIVITY

Building universal web associations could involve financial arrangements between providers. The understanding notes that such business dealings may include arrangements for the foundation and upkeep of providers offices [3].

INTERNET SECURITY IN E-COMMERCE

Nowadays, many more challenges are found which depend on an e-commerce website, for which security of the internet is essential. Since payments are conducted through the internet, cybersecurity is probably the major issue to consider. Frauds, hoaxes and schemes are not uncommon, and many are those that involve the theft of sensitive information or cash [4]. This would be called a bad client expertise, and for e-commerce organisations it is a reputation-tarnishing and probably business-sinking hurdle. There are some security protocols which are employed over the internet to affirm secured online transactions like secure sockets layer (SSL) and secure hypertext transfer protocol (SHTTP), etc. [4, 5].

SECURE SOCKETS LAYER (SSL)

Secure sockets layer (SSL) is a convention for PCs organising the relationship between customers and servers through a web as well as being like an unstable network. The use of SSL was rejected for web use by the Internet Technology Task Force (ITTF) [4] and replaced with the 'transportation layer security' (TLS) protocol because of different protocols, usage flaws and vulnerabilities, although TLS and SSL are not interoperable and TLS with SSL 3.0 is in reverse. It is the first convention that is regularly used and is commonly employed in the industry. It meets certain security criteria such as: verification, encryption and integrity [4, 5].

SSL could be a logical disciplinary convention giving a safe relationship between a customer and a server. SSL was [1–5] at first developed by the Netscape Communications Corporation. This convention protects secure data exchange between a customer and a server that allocates TCP/IP with an 'open key unbalanced calculation rule' utilised for the encryption strategy. The SSL arrangement includes two subcontracts: one for the SSL record and the other for the SSL handshake. The former defines the transmission location of the knowledge. The SSL convention incorporates a handshake with the SSL record convention to enable the flexible exchange of messages between a server and a customer.

The SSL convention builds up a protected channel with three fundamental features: i) all messages are territory encrypted, ii) authentication ought to be on server side as it is discretionary on the client side, iii) it gives guaranteed quality. Not only does SSL provide reliable information on the internet, but it also authenticates the server and customers. The SSL convention's basic aims are: i) cryptographic security – it offers a secure means of remotely exchanging data without any third-party impedance, ii) consistency – it is compatible with any stage, iii) expandability – new procedures might be implemented without any problem, iv) productivity – it transfers data through an open channel between two clients, or between server and client [5].

The SSL convention promotes three kinds of confirmation approval: customer and server validation, server verification with unauthenticated customer and full anonymity. A totally unknown meeting is, really, open to attacks. The TLS [5] convention was based on the possibility of a SSL 3.0 specification by the Netscape Corporation [6]. Its essential capacity is in providing well-being and information stability between two conveying applications. The essential tasks of TLS are much the same as those of SSL 3.0. There are two layers of TLS: the record protocol and the handshake protocol [1–6]. The first relies on anonymity and integrity of relationships, which are used to explain the more important conventions. A server and a client must each be determined by the handshake convention.

SSL Certificate Encompassing E-Commerce Transaction Security

SSL is a security standard to communication. An SSL certificate can be used as a digital file identifying your web server and encrypting the details you share with your users, in order to ensure that the client and the server have contact. TLS, which is typically synonymous with SSL, may also have been misunderstood. The certificates are the same and TLS literally is the technical term for the new versions of the protocol [5].

SSL Certificate Requirements

All exchanges sent over standard HTTP affiliations are plain content and can be interpreted by any programmer to prevent your software from associating with the web. This poses an unambiguous threat if the correspondence is arranged on request and contains a credit card or social security number. Every match is firmly scrambled with an HTTPS alliance. If someone has discovered how to tamper with the association, they would not be able to interpret any of the information that goes between you and the site [5].

Securing SSL Certificate

Users may request to share their identity when browsing a website containing such an SSL certificate. A duplicate copy of the SSL certificate is sent to the user by the server. The browser can now test the SSL certificate to determine whether it is confident. The client is aware of who gave the SSL certificate and who accepted or rejected the SSL certificate. If the certificate has been approved, the user may then start a secure session with the server or show an alert if the certificate has been refused. Now a customer can search your website safely and firmly and carry out transactions [6].

Digital Certificates

The digital certificate classification is composed of domain-validated certificates, extended-validated certificates, organisation-validated certificates and wildcard certificates.

a) **Domain validated certificates:** it goes with the benefit of providing the required validated certificates pertaining to the required domain. This makes domain SSL perfect for organisations requiring the permission of instant costs for SSL and without the effort of suffering organisation reports. Space-approved authentication offers low affirmation, because the organisation may not have approved these endorsements that do not encourage visitors to understand who is running the website. When you have a web-based company platform, the future customers are equally cautious. In some circumstances, it is used distinctly where visitors are not required to guarantee or wherever there is virtually no chance of external attack [6].

b) **Extended validated certificates:** it might be a kind of endorsement other than traditional SSL declarations which aims at preventing phishing attacks. A SSL certificate provider should do some complex approval to give one, it makes extended approved declarations uncommon [6]. The following verification ought to be done [5]:
   i.   Affirm that the association is legal and dynamic.
   ii. Affirm the location and number of the associations.
   iii. Affirm that the association has the option to utilise the area expressed in the EV SSL certificate [5].
   iv. Affirm that the association has endorsed the individual ordering.
   v.   Affirm that the association is not boycotted by any organisation [1–5]. All individuals from the 'declaration authority forum' should be prepared and set out in the referenced guidelines sought by all SSL providers [1–6].

c) **Organisation validated certificates:** in this case, to utilise a particular space called the 'endorsement authority' [3, 5] checks the privileges of the candidate. It handles the quality control of the association. The thoroughly checked organisation information is obvious to clients while tapping on the secure site seal and giving improved anonymity [7].

d) **Wildcard certificates:** it saves your money and time by making sure your area and unlimited sub-areas are on one authentication. Wildcard endorsements chip away at identical methods like a normal SSL certificate, allowing you to make sure the association between your site and your client's internet browser – with one significant concession is the wildcard SSL certificate – covers any and all of the sub-areas [8] of your principal webspace. It is easy to oversee and adapt as 'subject alternative names' (SANs) [1–6] made sure that wildcards might be a top option for associations dealing with different destinations facilitated across different sub-areas. With a digital certificate wildcard [6], you have the option to give duplicates of your endorsement on as few servers as you might want, everything about it is apportioned as its very own key, including to change or erase a SAN for any reason all through the lifecycle of the authentication for no further charge.

Importance of Security in Online Transactions

During online transactions, the user will share their personal data with their full name, address, mobile number and financial information such as credit and debit card numbers, user ID, password, etc. If the web transactions or the whole of the site are not secured with SSL-supported code, it will be a serious risk for users and easier for cybercriminals to attack and steal user information.

Ways to Secure Online Transactions

To make sure that online commerce is secure sites must offer character confirmation and encryption through a SSL certificate. SSL certificates offer durable encryption, configurable to suit the prerequisites of everyone from online shops to banks, to taxpayer supported organisations. The business ordinary 2048-piece keys and 'SHA-2' hashing calculations square measure are difficult for attackers to interfere with. To give more confidence to our visitors, we will use SSL certificates with 'organisation validation' and 'extended validation SSL certificate' that can work well to make e-commerce websites are secure

[5].

We can test the safety of the website by trying to find the following signals:

(i) **Address bar:** first search the address bar, when you see that your site begins HTTPS and the name of the commercial website is displayed in the address bar, it indicates that it is protected using SSL certificate validation. In addition, the name of the organisation that displays the website is secured with the highest degree of EV SSL certificate encryption if you discover a website address bar in green [6].

(ii) **Status bar color:**
   (a) The website is not trustworthy if the color of the status bar is red.
   (b) When the security status mark is yellow the certification authority does not check the authentication of the web site.
   (c) Once the security status is yellow, the certificate is authenticated and contacts between the user and the servers are encrypted.
   (d) If the security status is green, the EV SSL certificate secures the website, and consequently, the certificate's authority shall verify the identity of the site.

(iii) **Trust seal:** it is not just the EV SSL certificate, we may get the chance to check the trust seal on the site. The trust seal conveys data for that site such as domain name, address, contact details, SSL type, SSL expiration date and so on [6]; when you click on the trust seal you will get all the subtleties identified with e-business. The dangers of an unsecured e-commerce site [2–4] displays unbound site as well. Digital lawbreakers will essentially attack and steal clients' information. Attackers can do following things with an unsecured site:

a) Breach the personal and financial details of users.
b) Take advantage of malware on app and website systems.
c) Knowledge associated with breach of website.
d) Track the behavior of users.

According to the detailed examination of SSL certificates and e-commerce site security, you should introduce a SSL certificate on your server. It will make sure the related site, server, client data, exchange information and client's financial and sensitive information are secure.

SECURE HYPERTEXT TRANSFER PROTOCOL (SHTTP)

SHTTP is the scrambled revision to the HTTP convention, assembling a layer of security and compromising a site's susceptibility to assaults. Using SHTTPS gives assurance to clients and helps secure client data. HTTP adorns the 'HTTP protocol with open key encryption', confirmation and digital signature over the internet. The secure HTTP protocol fights various security components, giving security to the end-users. SHTTP operates by the execution of customer/server encryption. HTTP is anything but an independent convention, but a run of-the-mill correspondence convention working through TLS encryption instruments. HTTPS gives a more secure browsing experience which indicates that clients are not associating with a bogus site [5–8].

The https:// in the address bar and the design of the lock icon indicates an HTTPS affiliation, although it is not an failsafe guarantee you have the correct web address. Encrypted information protects login IDs, passwords, credit and debit card numbers and alternative personal information that is entered on a site. The encryption works irrespective of that direction and involves the data that are distributed. This point is of specific value to online traders and their buyers since it protects e-commerce transactions against thieves.

WORKING OF HTTP

HTTPS pages normally utilise all secure conventions to encode correspondence, SSL or TLS. Each of the TLS and SSL conventions utilise what is called an 'unbalanced public key infrastructure' (PKI) framework [5, 6]. A lopsided framework utilises two keys to encode correspondence [6, 7], an 'open key' and a 'private key' [7]. The non-open key is something that gets scratched with the open key.

As the names indicates, the private key must be guarded carefully and only the owner of the non-open key is available. The non-open key is securely tucked away on the online server inside a site case. On the other side, the encrypted key should not be circulated to anyone and they must be prepared to decode what was encoded with the non-open key [7].

On the contrary, HTTPS pages usually use one of two secure conventions for encoding, SSL or TLS interchanges. Both the TLS and SSL conventions use the PKI system called 'topsy-turvy'. An autopsy system uses two keys, an open key and a private key, to encode interchanges. Everything that is stored with the open key must be decoded with the private key [6].

SHTTP CERTIFICATE OVERVIEW

The site can send its SSL authentication to our program right from the beginning, when we ask SHTTP alliance for a website page. This will contain the open key to start the safe meeting. The SSL handshake begins to help this important exchange, your software and the site at that level [3–8]. The SSL statement includes shared insights, so that a guaranteed secure alliance can be established between you and the website. If a trustworthy SSL digital certificate is used all over the SHTTP, consumers can see a lock icon in the system address bar until the partner's 'enhanced validation certificate' is installed on an online website [5–7].

The site first sends its SSL approval for our programme when you ask for a SHTTP association with a page. This authentication requires the open key that the protected meeting is to begin. In this context, your programme and site start the SSL handshake. In view of this underlying business, the SSL handshake requires common insider evidences to create a secure relationship between you and the website. The clients see a lock icon in the site's address bar when a trusted SSL digital certificate is used in a SHTTP association. The address bar will become green when an extended validation certificate is displayed on the web [6, 7]. In order to make a web server HTTPS, the web server should make a declaration.

Validation of HTTPS uses two declaration styles as provided by the Comptoir Electrique Aquitain (CEA). (i) Server declarations: these the keys by which coding is implied. Essentially, they are the figurative content segments that ensure secure connection among the customers. (ii) Client declarations: this database guarantees that the non-open information about a client guarantees that an SSL customer's ID is transmitted to a server. You will use your own declarations without the certificate authority. This is possible. They are referred to as authentications marked by themselves, restricting them to being weaker than servers and buyers [2, 5, 7, 9].

A web server must set up an endorsement for that web server in order for it to process HTTPS. Verification of HTTPS uses the certificate authorities for two kinds of endorsement:

(i) Server authentications: these are the keys provided by encryption methods. They are the content components used to ensure a secure connection between the collection of customers.

(ii) Client declarations: individual client data are conveyed within this approved assurance, which offers distinctive proof of an SSL customer from a server. In relation to the certification authority, you can make your own declarations. They are known as self-marked declarations and viewed as less secure compared with server and customer ones [5–8].

A)   **Server Side:** a server declaration is to be issued to the shopper in order to begin, before setting up an SSL connection to a client. The server will issue a certification authority in the name of NetBIOS or DNS [4, 6, 7]. A certification authority that guarantees that the maker of the declaration has been checked will include the server endorsement containing the certification authority signature, the total population key of the sender, additional details from a few collectors and a validity date. Each statement also contains two key elements: open key and private key. In this way, a non-open key is used to decode the traffic received from the customer to validate it, which means that it is necessary for the two-way trust relationship to be formed by a marked authentication [6].

In order to get started, the client must be issued a server certificate before creating an SSL connection to the network. A server certificate may be issued by a NetBIOS or DNS certification authority. A certification authority that guarantees the certificate owner's verification will provide a server certificate that includes the certification authority signature, a public sender key, some additional information about the recipient and the validity date [5].

There are two keys to each certificate: one private and the other public. The public key helps to encrypt and decrypt the data accessed by the client using the private key. It is necessary to demonstrate the trust relationship in two ways through a signed certificate to perform the authentication validity [6].

B) **Client Side**: the next half is to be with the consumer with each of the things done on the server. The service certificate is issued by the certification authority. In general, the mapping of customer certificates can be a basis for validating authentication. If all the conditions set by server are fulfilled, the sending (SSL client) or the receiving (SSL server) must generate and exchange SSL session keys [3].

The prime advantages of a SHTTP certificate are:

(i)   Customer information, like a MasterCard number, is encrypted and cannot be accessed.
(ii) Visitors will check that you are a registered company that owns the domain.
(iii) Users trust and complete purchases from sites that use HTTPS.

The customer certificate is issued by the certification authority. Mapping client certificates is usually a basis for validating the authentication. A sender (SSL client) and a receiver (SSL server) build and share SSL session keys if all the conditions set by the server are met [4–6].

SEO EDGES

a) Enhanced rankings

Google announced in 2014 that websites fitted with the shopping protocols would earn a slight ranking increase over HTTP websites. For Google, this can be just a soft signal. If the same technical and content links were given to both sites, Google would rank web pages with the shopper protocols via HTTP websites. For the future, this may be a stronger signal.

b) Insecure login

The introduction of Chrome 57 will provide users with informed security alerts on sites that often run the protocol for purchasers at the lowest level. In early 2017, Chrome additionally gave users a 'not secure' warning for any protocol websites that asked for login or MasterCard info. Imagine, however, whether users can understand conducting a Master Card transfer on your website if their browser tells them it is not secure.

c) Referral data

Traffic passing through the AN protocol server appeared to be a direct traffic once the maltreatment analytical software system does its operation.

d) Security

HTTPS ensures the right web page on the server is secure. In order to defend against breaches of third-party rules, HTTPS additionally encrypts all user data as well as financial information and browsing history.

e) Quick browsing

Many browsers enhance the HTTPS convention, these give programme upgrades over the standard convention. Once HTTPS is enabled, users can experience faster browsing speeds, similar to encoding. Major enhancements to transport layer security (TSL) have made data more secure.

f) Mobile

With the latest mobile version of Google, the websites are encouraged to be converted into HTTPS, which may impact rankings more than desktop searches. Google requires that websites be equipped with SSL in order to convert web pages to AMP, which could dramatically influence organic mobile rankings. Many websites have struggled to receive HTTPS certificates because of financial difficulties. One reason why Google has increased the rankings of HTTPS client websites is the long process of obtaining an 'extended validation' (EV) certificate. Websites which offer many of the same cryptography and security advantages as the EV certificates offered free and automated 'domain validation' (DV) certificates.

Websites such as WordPress upgrade websites automatically for HTTPS, and business leaders such as Amazon give their customers TSL certificates. A more streamlined and easier-to-operate HTTPS security certificate has been activated.

Implementing a broken HTTPS link is not enough because it undermines user trust and jeopardises your website. Qualys SSL Labs offers testing tools for TSL connection configuration. An HTTPS connection can frequently make data for third parties like JavaScript and security-inflicted images. 'HSTS' can be used to fix content problems causing security warnings like 'HTTP'. Ultimately, switching to HTTPS will rely on your company goals and on your digital marketing strategy. If our website does not provide login details or funding transactions, it is best to avoid a future problem or create a broken HTTPS link. Read the best practices of Google for HTTPS implementation before deciding to make this switch.

SECURE DIGITAL TRANSACTIONS

This protocol was proposed with MasterCard and Visa in collaboration with others. In theory it is the best defence protocol. The following components are included:

- **Digital wallet software:** it lets the owner of the card point and click to GUI to make secure online transactions.
- **Merchant software**: this app lets retailers connect efficiently with prospective clients and financial institutions.
- **Payment gateway server software:** both standard and automatic payment systems are available via the payment gateway. It supports the process of applying for certificates from merchants.
- **Certificate authority software:** such programmes are used by financial services providers to produce cardholders' and merchants' digital certificates and enable them to enter their electronic business account protected agreements [11].

Various Methodologies to Protect E-Commerce Sites from Hacking

Several techniques are used repeatedly for providing security to e-commerce sites from severe hacking. The following are few of the techniques used for this purpose:

- Choosing a stable platform for e-commerce.
- Employing a secure online checkout connection and ensuring that you are PCI compliant.
- Not storing sensitive information.
- Employing an address and card verification framework.
- Requiring strong passwords.
- Setting up system alerts for suspicious activity.
- Using tracking numbers for all the corresponding orders.
- Tracking your site regularly and check whoever is hosting.
- Patching your systems.
- Ensuring DDoS security and mitigation tools are available.
- Drawing up a disaster recovery plan to make sure that you or someone on your site knows how to implement it.

MAJOR STRATEGIES EMPLOYED TO CURTAIL FRAUD IN E-COMMERCE

The following are the widely used popular approaches used to combat e-commerce fraud [8]:

A) **Risk-Based Authentication (RBA)**

   As the cost of a fake exchange is commonly recouped after the exchange authorised, the RBA procedure endeavors to trap false transactions as they occur. At the core of this technique is a variety of selected devices that filter through a lot of exchange information revealing shopping practices, transnational practices and more in a way to deal with structuring an ongoing assurance as to whether a particular exchange ought to be confirmed or not [12].

B) **Data Tokens**

   Cardholder security is already hindered by the number of information stores expected to empower electronic exchanges. With the idea of a data token, the basic cardholder ID information is made very secure. The main applications getting access to the real cardholder's information, for example, name, social security number, account number and so on, are those fundamentals for the exchange procedure. Each and every other component in the exchange chain gets an information token, an arbitrary distinguishing proof code produced by the framework, which has an incentive to anybody with the exception of dishonest traders. By the use of information tokens, organisations secure such sensitive information at the source, while decreasing their expense of ensuring such information at the same time as meeting the information protection guideline [13].

C) **Brand Security**

   In this methodology, the brand plays a role against misrepresentation and works to ensure the brand. Practices to this end are not restricted to examining the URLs, actualising every minute of every day of web activity, blocking recently known bogus sites and alerting clients soon after an issue is discovered. Most purchasers feel that it is the brand's responsibility to protect itself from misrepresentation. A well-executed phishing methodology does substantially more than swindle purchasers. It fundamentally undermines the assaulted brand which will get accused of paying little attention to its responsibilities [14].

PAYMENT SYSTEMS IN E-COMMERCE

An online corporate payment network allows electronic transfers to be acknowledged for online exchanges. Such exchanges are likewise viewed as a test of 'electronic data interchange' (EDI) [8–10]. Different online transfer approaches are as per the following:

a) **Payment through bank:** such a framework does not involve any kind of physical card. This is utilised by clients who have internet banking accounts. Rather than entering card details on the buyer's site, in this framework the transfer process permits one to determine which bank they wish to pay from [9]. At that point the client is diverted to the bank's site, where they can verify themself and afterwards confirm the payment. Commonly there will likewise be some type of two-factor verification. A few providers, such as Trustly, let financial service providers insert i-frame on their site so that the purchasers can pay without being diverted away from the first site [10, 15].

b) **Payment through PayPal:** it is a worldwide online business permitting payments and cash moves to be made through the internet. Online transactions are processed as electronic options in contrast to paying with conventional paper techniques, for example, cheques and cash orders. It is dependent upon the US monetary authorisation list and different principles and interventions required by US laws or government. PayPal is an acquirer, performing payments for online merchants, sells off destinations and other business clients, for which it charges a fee. It might likewise charge a fee for withdrawing cash, relative to the amount withdrawn [9, 10]. The fees are related to the cash utilised, the transfer method utilised, the nation of the sender, the nation of the beneficiary, the sum sent and the beneficiary's account type. Also, eBay purchases made with a chargecard through PayPal may incur additional fees if the purchaser and vendor use various monetary forms. On October 3, 2002, PayPal became a wholly owned subsidiary of eBay [16].

c) **Payment through Paymentwall:** Paymentwall, an internet business platform founded in 2010, offers a wide range of online payment methods that its customers can use on their sites [9, 10].

d) **Payment through Google Pay:** Google Wallet (now Google Pay) was launched in 2011, serving a similar role as PayPal to encourage transfers and move cash on the web. It likewise includes a security that has not been broken to date and the capacity to send transfers as connections by means of email [9, 10].

e) **Payment through mobile money wallets:** in developing countries numerous individuals do not use bank branches, particularly in level II and level III urban communities. Taking the case of India, there are more cell phone clients than there are individuals with active financial balances. Telecom providers in such areas, have begun offering portable cash wallets which permit holding balances, effectively through their current mobile number, by visiting physical agents near their homes and workplaces and changing their money into versatile wallet money. This can be utilised for online exchange and e-commerce purchases. Numerous alternatives, for example, Airtel Money and M-Pesa in Kenya, Access to Work (ATW) is being acknowledged as substitute payment choices on different e-commerce sites [9, 10].

CONCLUSION

Basically, a web-based business is for the purchase and sale by organisations and shoppers of goods and services on the internet. People use the electronic business or web shopping to search for and purchase things online and a short time after searching use a charge card. Web bargains are growing rapidly as buyers: take advantage of the lower costs offered by venders working with fewer overheads than a physical store, enjoy the convenience of having items delivered instead of incurring the cost of time and transport of going to a store, source cheaper things from overseas vendors, explore the exceptional variety and stock offered by online stores, use search engines that filter and suggest items and sale offers where applicable.

REFERENCES

1. Begam, S., Selvachandran, G., Ngan, T. T. & Sharma, R. (2020). Similarity measure of lattice ordered multi-fuzzy soft sets based on set theoretic approach and its application in decision making. *Mathematics, 8*(8), 1255.
2. Thanh, V., Sharma, R., Kumar, R., Son, L. H., Pham, T. et al. (2020). Crime rate detection using social media of different crime locations and twitter part-of-speech tagger with brown clustering, *Journal of Intelligent and Fuzzy Systems*, (pp. 4287–4299).
3. Nguyen, P. T., Ha, D. H., Avand, M., Jaafari, A., Nguyen, H. D., Al-Ansari, N., ... Pham, B. T. (2020). Soft computing ensemble models based on logistic regression for groundwater potential mapping. *Applied Sciences, 10*(7), 2469.
4. Jha, S., Kumar, R., Hoang Son, L., Abdel-Basset, M., Priyadarshini, I., Sharma, R. & Viet Long, H. (2019). Deep learning approach for software maintainability metrics prediction. *IEEE Access, 7*, 61840–61855.
5. Sharma, R., Kumar, R., Sharma, D. K., Le Hoang, S., Priyadarshini, I. & Pham, B. T. (2019). Dieu Tien Bui & Sakshi rai. Inferring air pollution from air quality index by different geographical areas: Case study in India. *Air Quality, Atmosphere and Health, 12*(11), 1347–1357.
6. Sharma, R., Kumar, R., Singh, P. K., Raboaca, M. S. & Felseghi, R.-A. (2020). A systematic study on the analysis of the emission of CO, CO_2 and HC for four-wheelers and its impact on the sustainable ecosystem. *Sustainability, 12*(17), 6707.
7. Sharma, S., Kumar, R., Das Adhikari, J., Mohapatra, M., Sharma, R., Priyadarshini, I. & Le, D. N. (2020). Global Forecasting Confirmed and Fatal Cases of COVID-19 Outbreak Using Autoregressive Integrated Moving Average Model, ront. *Public Health*. doi: 10.3389/fpubh.2020.580327.
8. Malik, P. et al. (2021). *Industrial Internet of things and its applications in Industry 4.0: State-of the art, computer communication* (Vol. 166, pp. 125–139). Elsevier.
9. Sharma, R., Kumar, R., Satapathy, S. C., Al-Ansari, N., Singh, K. K., Mahapatra, R. P., ... Pham, B. T. (2020). Analysis of water pollution using different physicochemical parameters: A study of Yamuna River. *Frontiers in Environmental Science, 8*. doi: 10.3389/fenvs.2020.581591, pp. 581–591.
10. Dansana, D., Kumar, R., Parida, A., Sharma, R., Adhikari, J. D., Van Le, H., ... Pradhan, B. (2021). Using susceptible-exposed-infectious-recovered model to forecast coronavirus outbreak. *Computers, Materials and Continua, 67*(2), 1595–1612.
11. Vo, M. T., Vo, A. H., Nguyen, T., Sharma, R. & Le, T. (2021). Dealing with the class imbalance problem in the detection of fake job descriptions. *Computers, Materials and Continua, 68*(1), 521–535.
12. Sachan, S., Sharma, R. & Sehgal, A. (2021).Energy efficient scheme for better connectivity in sustainable mobile wireless sensor networks. *Sustainable Computing: Informatics and Systems, 30*. PubMed: 100504.
13. Ghanem, S., Kanungo, P., Panda, G., Satapathy, S. C. & Sharma, R. (2021). Lane detection under artificial colored light in tunnels and on highways: An IoT-based

framework for smart city infrastructure. *Complex and Intelligent Systems*. doi: 10.1007/s40747-021-00381-2.

14. Sachan, S., Sharma, R. & Sehgal, A. (2021). SINR based energy optimization schemes for 5G vehicular sensor networks. *Wireless Pers. Commun.*. doi: 10.1007/s11277-021-08561-6.

15. Priyadarshini, I., Mohanty, P., Kumar, R., Sharma, R., Puri, V. & Singh, P. K. (2021). A study on the sentiments and psychology of twitter users during COVID-19 lockdown period. *Multimed Tools Appl.*. doi: 10.1007/s11042-021-11004-w

16. Azad, C., Bhushan, B., Sharma, R., Shankar, A., Singh, K. K. & Khamparia, A. (2021). Prediction model using SMOTE, genetic algorithm and decision tree (PMSGD) for classification of diabetes mellitus. *Multimedia Systems*. doi: 10.1007/s00530-021-00817-2

9 Identification of Paddy Diseases by Various Algorithms–A Review

S. Ramesh and M. Vinoth Kumar

CONTENTS

DOI: 10.1201/9781003097518-9

INTRODUCTION

Agriculture becomes much more important with a growing population. There are several diseases affecting rice plants such as bacteria, fungi or viruses. These diseases are subdivided into various types. Each disease is mentioned. The bacterial diseases are bacterial leaf streak, bacterial blight, grain rot, root rot and sheath brown spot. The fungal diseases are black kernel, brown spot, aggregate sheath spot, brown sheath rot, false smut and blast (leaf, neck, nodal and collar). According to the report by the International Rice Research Institute (IRRI) rice diseases occur in 85 countries. Disease will affect the rice crop, which causes damage at the nursery stage and main field stage. The first diseases in rice crops were found in China in 1637; at that time these diseases were called rice fever diseases. In India, rice diseases are usually called rice fever or rotten neck. In the rice plant, blast is the major disease; it will affect all parts of the plants and cause loss of up to 100%, and the expected grain loss in India due to this blast disease is 70–80% and in Philippines is 50–85%. The blast diseases were first recorded in India 1918 and in China (1980–1981), Korea (mid-1970s). It is possible to detect rice diseases in different ways. Some of the diseases are not visible to the human, eye in which case it is very late to act. Remote-sensing techniques are used to survey multi- and hyper-spectral images. The next section describes the literature survey, which is followed by a section on detection. Further sections deal with classification, and quantification, and these are followed by a presentation of the discussions and a conclusion.

LITERATURE REVIEW

Agriculture is the origin of human settlement on Earth. Development will depend upon agricultural production. As per the report, in our country 70% of the people include rice in their daily food. Every year during the cultivation season 37% of the crops are affected by disease. Diseases can affect all parts of the plants and it are sub divided into various types. The authors classify rice diseases using different techniques which include detection, quantification and classification. All the methods are explained in this chapter. More information on the subject is available in the paper by Jayme Garcia Arnal Barbedo (2013).

DETECTION

The prime focus of this section is to find out the severity of the diseases. Image processing techniques are used to establish the detection algorithms. The methods of the image processing are that images are acquired from the field and pre-processing will be done, then enhancement of image followed by extraction of features and classification.

IMAGE PROCESSING METHODS

Pujari et al. (2015) explain the study of image-processing techniques to recognise and classify fungal disease symptoms on horticultural and agricultural crops. The aim is to identify, quantify and detect the initial symptoms of diseases.

Devi et al. (2014) explained the work of the segmentation algorithm, to identify the affected part of paddy leaves. In the image segmentation method, if there is varying grey level background a problem has occurred. Various segmentation techniques are applied, such as work mean shift segmentation and region growing segmentation. The quality of the segmentation is improved by using color space transformation and median filters.

Artificial Neural Networks

Orillo et al. (2014) proposed the identification by manual inspection of three common diseases by digital image processing of brown spot, bacterial leaf blight and rice blast. By image processing techniques four features were extracted. For the best performance and accuracy, a back propagation neural network was used. Some 134 disease-affected images were taken, 70% were used for training the network, 15% for testing and 15% for verification. After completion of the processing images, detection of diseases was 100%accurate.

Phadikar et al. (2008) proposed a prototype software system to detect rice diseases. Using a digital camera, images of the infected parts were captured. The diseased portion of the plant was analysed by image segmentation methods. For classification purposes the self-organising map (SOM) and neural network was used.

Mutalib et al. (2016) proposed image processing methods in the paddy field. The paper focused on system architecture to detect the diseases for the first time. The work was implemented with the back-propagation neural network (BPNN).The BPNN is divided into three layers which consist of input layer, hidden layer and output layer. In each layer the nodes are identified as 2500, 1250 and 3 respectively. The output layer represents the three types of disease such as leaf blast, bacterial leaf blight and sheath blast. The affected images are collected in the early and middle stages. For leaf blast, 33 images were collected in the early stage and 43 images in the middle stage. For bacterial leaf blight 28 images were collected in the early stage and 28 the middle stage. For sheath blast 32 images were collected. A total of 164images were involved in this consideration work. The ratio of the images was 80:20, 80% (133 images) was used for training purposes and 20% (31 images) for testing. Finally, accuracy of 70 to 80% was achieved.

Jingbo et al. (2009) proposed a machine-learning approach to find out the protein in rice for disease resistance. The method was assessed using artificial neural networks. Some30 features were taken for the work and the result obtained was 92.86%. After the features process, a feature-reduced classifier was used for different tests, the results of the test were Matthews correlation 0.4419; resubstitution test 100% and jack knife test 72.13%.

Knowledge Based

Prasad et al. (2014) proposed to obtain disease images in real time; mobile phones were used. The technique used here is threshold-based offloading. The algorithm is optimised for various Android-based mobile devices. CIE L*a*b color space is used as it is more flexible and device independent. To get the unsupervised image segmentation, a camera was used to capture the images using various resolutions on several mobile phones to reduce the communication cost in real time while transmitting the images and to reduce the power consumption. Finally, the result was that the total power consumption was reduced while using different mobile phones.

Mercy Nesa Rani et al. (2013) designed the concept of an expert system to identify rice diseases. The diseases detected were false smut, brown spot, sheath blight blast and sheath rot. A set of rules was developed with the help of the system. In the database, diseases were uploaded and stored in the farmers specific languages, so the farmers can easily access the system. When any type of new disease occurred they could upload the details to the database so they could be observed by the domain experts and they could get information about the new diseases from them. The primary research of this proposed method was to detect the disease in the initial stage to save the crops and increase the productivity for farmers.

Robindro et al. (2013) described an expert system for disease identification. The system was developed with Java expert system shell (JESS).A knowledge-based system contains a rule-based system which will follow the symptoms of the diseases and stop them spreading to other plants. Finally, the result was that it is not applicable for rice diseases alone as it will show various diseases symptoms using artificial intelligence.

Fuzzy Logic

The method of digital image processing and machine vision technology proposed by Rastogi et al. (2015) is used for leaf disease identification and grading. The proposed system consists of two phases: phase one is recognition of the plant by leaf features, it includes leaf images for pre-processing, feature extraction and artificial neural network (ANN) classification for training of the leaf recognition. In the second phase the affected leaf area is classified by k-means-based segmentation, feature extraction of defect portions is used to find out the disease by ANN-based classification. The grading is used to find out the number of diseases available in the leaf.

Leaf diseases can be detected automatically in plants by using machine vision technology and ANNs. The segmentation of the image can be done by Euclidean distance techniques and k-means clustering techniques to segment the input leaf image of the disease area and to calculate the infection percentage of diseases on the leaf, and then categorise them into different classes.

Classification

Classification methods are an extension of the detection methods, here it is explained as instead of trying to detect different conditions and symptoms for only one specific disease. This will be useful for the researchers to identify and analyse which disease affects the plant. The author explains the classification methods by various algorithms.

Remote Sensing

A method was proposed by Santanu et al. (2016) to identify rice blast disease by automatic identification. From the normal portion of the plant, the infected region can be easily detected by an automatic identification system. The greenness of the remote sensing images can be found using a vegetation index (VI). Four types of VI are used to get the acquired images of the infected plans and which include NDVI, GNDVI, EVI and SAVI. The images of the infected rice leaf (leaf blast and brown spot) were taken into consideration. To get efficient segmentation, a VI is used. To extract the VI images Otsu's method is applied. Five different texture features (homogeneity, contrast, correlation, energy and entropy) were computed in infected regions. To classify the diseases the author used 15 different classifier feature values. From the four types of VI two gave the best results (VI, EVI). The classification method was analysed by two methods which included the proposed method and the conventional method. The accuracy was 84% for the proposed method on detecting the two diseases and 82% for the conventional methods.

Phadikar et al. (2012) proposed that pre-processing of the image analysis is used for eliminating the shadows from the acquired images. Novel shadow removal techniques have been introduced in the rice fields to eliminate shadows from the images. To remove the shadows of the agriculture field from image the VI, RI and NDGI of the images are calculated. The edges of the images can be found by the canny edge detection algorithm, to eliminate the noisy edges a median filter is applied. Image processing operations will affect the performance of the shadows of the acquired images. To remove the shadow from the images, the shadow elimination method uses VIs.

Membership Function

Anthonys et al. (2009) proposed the recognition of rice blast, brown spot and rice sheath blight paddy diseases in the agricultural field. To segment these images mathematics morphology was used. A color image of the leaf with disease spot was extracted by its color features, texture and shape. The three types of diseases were determined by the membership function of the classification method. From 50 extracted images, 70% accuracy was obtained. Finally, the recognition time of the rate was less than 2 seconds.

Feature-Based Rules

Kurniawati et al. (2009) designed the concept of a recognition system to identify the narrow brown spot, brown spot and blast rice diseases. In the classification task five features were tested which included broken paddy leaf color, lesion type, boundary color, lesion percentage and spot color. To get the best result from the 72 images three thresholding techniques were applied. With the three methods, the best accuracy was given by variable threshold which was about 86%.

Two thresholding methods were used for the 94 paddy leaf images in another publication that same year to get the best results. The accuracy was estimated to be around 94.7 percent.

Neural Networks

Sanyal et al. (2007) proposed that detection of a balance deficiency in the mineral level was done by image analysis techniques. From the total, 80% of images were taken for training purposes and 20% for testing. The image pixel considered a 7 × 7 neighborhood for computation of the feature vector. The multilayer perceptron (MLP) classifier of texture features and color features consisted of 40 hidden holes and 70 hidden holes respectively. The results of the correctly classified pixels were 88.56%.

Liu et al. (2009) proposed that healthy and diseased parts of rice leaves are classified using BP neural network classifiers. Rice brown spot was taken for the work; training and testing samples were collected from the northern part of the Ningxia Hui Autonomous Region. The results were realistic to identify rice brown spot by image analysis and BPNNs. Some 20 pixels were selected for training purposes which included lesion color input as 10 samples and background color input as ten samples. The result showed that the color features of the brown spot region were good features which differed from the healthy region. Finally, it was stated that this method is not only for brown spot disease; it is suitable for other types of diseases as well.

Zhou et al. (2010) proposed BPNN to identify rice disease spot with image segmentation. The common disease tested in this work was rice blast. For BPNN, the mixture of divergent color features parameters determined the input first. Second, to construct the rice blast spots segmentation included five as inputs, ten as hidden neurons and one as output. The results obtained from the test were 99.8% accurate.

Asfarian et al. (2013) proposed that for the four major diseases (leaf blast, bacterial leaf blight, brown spot and tungro) the texture of the lesions could be analysed by fractal descriptors. The classification process was used for each lesion by Probabilistic Neural Networks (PNN).There was 83% accuracy in identifying the diseases with these techniques.

Majid et al. (2013) proposed a paddy disease identification system for mobile application using fuzzy entropy and PNN which runs on the Android mobile operating system. The reason to select paddy disease was that yield loss affects profit as well. Paddy diseases were extracted using a digital paddy leaf and classified under PNN. The result shown in the identification of paddy diseases was 91.46% accurate.

SUPPORT VECTOR MACHINE (SVM)

The method proposed by Yao et al. (2009) to detect rice diseases in the early stage used image processing techniques and SVM. The accuracy (97.2%) showed that the SVM will detect the disease spots effectively. From each disease, 72 disease spots were randomly divided into two groups in the ratio 1:1. The classification models were developed into three: first as four shape and 60 texture features, second as 60 texture features and third as four shape features. RBLB, RSB and RB will represent the models in order as 1, 2 and 3. The radial basis kernel was selected to classify the diseases by SVM. Model 2 dropped accuracy to 88% due to the performance of similar textures between RSB and RBLB. Model 3 used only shape features so the accuracy dropped to 82.4%. The RBLB structure was lengthy so it easily differed from RB and RSB, it could not classify the other two diseases. Finally, the results revealed that from the three diseases, using shape and texture features will attain accuracy in high classification.

Ren et al. (2010) proposed the disease resistance of the plant. The gene expression of the data divided into two types: small samples and high dimensions. The results of the four genes were placed as first, third, sixth and eighth respectively. Strong relationships in the resistance of the diseases were shown by the first two genes, the next two genes did not show resistance to the diseases due to impact of an irritation response in the plants. Out of eight resistances in the plants, four of them had a certain effect on plant disease resistance.

REMOTE SENSING

Zhihaoqin et al. (2003), proposed that the purpose of remote sensing is to find rice disease detection in visible and infra-red regions. Correlations were applied between the ground data and imagery data. It indicated that broadband remote sensing imagery is suitable for this application. Some images had a correlation of 0.62(RI14, SDI14 and SD124).It was valuable for rice disease identification. By this application ground data had accuracy of 70% by classification. The error rate in the method was nearly 13%, due to heavily overlapping images it was hard to differentiate the healthy plants from the infected ones (DI < 20%). Identification was more accurate when the infection region was from mid to severe levels (DI < 35%). Finally, it was concluded that the remote sensing broadband imagery was suitable for rice disease detection systems.

QUANTIFICATION

The methods proposed in this system are to find out the quantity and seriousness of the diseases. The different techniques used for quantification are explained below. The main purpose of quantification is to estimate the severity of the diseases.

FUZZY LOGIC

Zhiyanzhou et al. (2009) proposed that rice plant-hopper (RPH) (*Nilaparvatalugens*, *Laodelphaxstriatellus* and *Sogatellafurcifera*) infestation is considered a more serious diseases in Asia. Visible images are needed to detect the stress in rice production by RPH infestation, the algorithms have been developed (fractal eigenvalues and fuzzy C-means), and images of rice stems have collected to start the experiment design. The visible images were pre-treated with image color space, denoising, smoothing and frequency domain transformation. The images of eigenvalues were extracted using the FCM clustering algorithm. From the RPH infestation the classification and detection were done by FCM clustering algorithms and fractal eigenvalues. The accuracy obtained was 87%.

NEURAL NETWORKS

Kun et al. (2012) proposed the rice blast grey system with two models: grey ant colony GM (1, 1) and Radial Basis Function (RBF) neural networks, combined and presented these two models. Though the research focused on the rice blast of the grey system, the following models were used as colony prediction, combination prediction was used to obtain the accuracy of RBF neural networks. By following this model, any model or combination prediction model can be analysed and predicted. Accuracy of 96.77% was obtained from the combination prediction model when compared with the other two models. Finally, the work suggested that the techniques were suitable for rice blast disease prediction.

EXPERT SYSTEM

Balleda et al. (2014) stated that for developing countries like India, agriculture is the main origin of yield with crops such as rice and wheat. Pests will degrade the quantity and quality of the crops. For pest management the author designed the rule-based expert system in architecture framework. The system was designed with different modules, was easily accessible and to know about the disease symptoms. The explanation block (EB) will provide the explanation of the system when a decision has to be taken; the EB will follow rules of the kernel in the expert system. Agpest is taken as the input and is tested for all the cases. Results of rice module and wheat module are listed in this paper tables (Table III and IV) and showed the complete statics of the Agpest. The speed and time are performed mainly by these listed two tables. Table 9.1 shows that the time taken for process decisions. Finally, the expert system is suitable for timely detection of diseases in accurate time.

Kalita et al. (2016) proposed a knowledge-based expert system to recognise diseases symptoms in the crops. The disease symptoms taken were leaf blast and neck blast. A set of rules was designed to follow the system. The system was designed and developed in various phases. In the first phase knowledge was acquired from books, literature and various sites. The acquired knowledge was arranged in categories and set of rules applied to them. For identifying a particular disease the forward chaining mechanism was used. The user can communicate their diseases and match with the database, as the system was created to be user friendly. The expert system will identify the disease in real time and send the information to the end user. All the acquired images are stored in a structured format. Users can come and discuss the diseases with the remote users throughout their lifespan.

SVM

Joshi et al. (2016) described the idea of image processing techniques to monitor and control rice diseases. The diseases taken for the work were brown spot blast, bacterial blight and sheath rot. To test performance, 115 leaf images were evaluated. For

TABLE 9.1
Detection

Reference	Main Tool/ Algorithms	Rice Crop Diseases	Salient Features
Pujari et al. (2015)	Image processing	Fungal diseases	The outputs of the feature extraction are stored in the database
Devi et al. (2014)	Image processing-based segmentation	Diseased portion of the paddy leaves compared for segmentation	To segment the rice leaf diseases the performance is obtained by means a shift algorithm and law color space transformation
Orillo et al. (2014)	Neural networks	Rice blast, bacterial leaf blight and brownspot	After completion processing of the images, diseases aredetected with 100% accuracy
Phadikar et al. (2008)	Self-organising maps (SOM)	Brown spot, leaf blast	The successful classification percentages are: Sample1 92% Sample2 84% Sample 3 82% Sample 4 70%
Mutalib et al. (2016))	Back-propagation neural networks (BPNN)	Sheath blast, bacterial leaf and leaf blast	Finally,accuracy of 70–80% has been recorded
Jingbo et al. (2009)	Artificial neural networks	The protein in rice for the diseases that are resistant or not	30 features are taken for the work and the result obtained is 92.86%
Prasad et al. (2014)	Knowledge based	Plant disease diagnosis	Total power consumption has reduced the mobile devices
Mercy Nesa Rani et al. (2013)	Knowledge-based system	Brown spot sheath rot, false smut, blast and sheath blight	Identify the disease in very advanced stage and the crop could be saved
Robindro et al. (2013)	JESS (Java expert system shell)	Disease diagnosis	Knowledge base contains rule-based system which will follow the symptoms of the diseases and stop spreading to other plants
Rastogi et al. (2015)	Fuzzy logic	Leaf diseases identification	From the disease leaf,the percentage of infection is calculated and categorisedinto several classes

classifier training 70% of images were used and for 30% for testing. The diseases were classified using two different classifiers: minimum distance classifier (MDC) and k-nearest neighbor classifier. The overall accuracy obtained from the kNN classifier was87.02% and MDC classifier was 89.23%.

Tables 9.1–9.3 differentiate the summarisation of the proposals.

TABLE 9.2
Classification

Reference	Main Tool/ Algorithms	Rice Crop Diseases	Salient Features
Phadikar et al. (2016)	Remote sensing-vegetative indices	Brown spot and leaf blast	While detecting two diseases 84% accuracy occurs and conventional method accuracy is82%
Phadikar et al. (2012)	VI and edge textures	Brown spot, sheath rot	To eliminate the shadows from acquired images
Anthonys et al. (2009)	Membership function	Brown spot, rice blast and rice sheath blight	Recognition time rate is obtained within less than 2 seconds
Kurniawatiet al. (2009)	Feature-based rules	Brown spot, blast and narrow brown spot	94.7% of the accuracy obtained by entropy threshold of the two methods
Kurniawati et al. (2009)	Feature-based rules	Brown spot, blast and narrow brown spot	86% of the accuracy obtained by variable threshold of the three methods
Sanya et al. (2007)	Neural networks	Mineral level deficiency	88.56% of the pixels are exactly classified by the quantitative analysis
Liu et al. (2009)	Neural networks	Leaf diseases	The recognition rate is more than 90%
Zhou et al. (2010)	Neural networks	Rice blast	From the two methods, accuracy of the test obtained is 99.08%. The segmentation accuracy rates of improved BPNN and standard BPNN gives 84.8% and 84.9% respectively.
Asfarian et al. (2013)	Neural networks (BPNN)	Tungro, bacterial leaf blight, leaf blast and brown spot	The overall accuracy is 91.80% and individual classes accuracy are: brown spot 92.31%, leaf blast 83.00%, bacterial leaf blight 96.25%,tungro 97.96%
Majid et al. (2013)	Neural networks (PNN)	Paddy diseases	The paddy disease identification accuracy is 91.46%
Yao et al. (2009)	Support vector machine (SVM)	Rice sheath blight, bacterial leaf blight and blast	The final classification accuracies are: Model 1 is 97.2%, Model 2 dropped to 88.0%, Model 3 dropped to 82.4%
Ren et al. (2010)	SVM-RFE	Rice genes	Using SVM-RFE the rice gene expression data can be classified for some level of success

TABLE 9.3
Quantification

Reference	Main Tool/ Algorithms	Rice Crop Diseases	Salient Features
Zhou et al. (2011)	Fuzzy logic	Rice planthopper (RPH) (*Nilaparvatalugens*, *Laodelphaxstriatellus*), and *Sogatellafurcifera*.	When differentiate into four groups accuracy is 63.5%
Kun et al. (2012)	Neural networks	Rice blast	The combination prediction model accuracy is high, up to 96.77%
Balleda et al. (2014)	Expert system (knowledge based)	Disease diagnosis and pest controls	Table III shows the results of AgPest of the rice module
Kalita et al. (2016)	Knowledge based	Neck blast and leaf blast	Particular disease has been identified using forward chaining mechanism
Joshi et al. (2016)	SVM	Rice bacterial blight (RBB), rice blast (RB), rice brown spot (RBS), rice sheath rot (RSR)	The overall accuracy obtained from kNN classifier is 87.02% and MDC classifier is 89.23%

DISCUSSION

Tables 9.1–9.3 show the technique with different algorithms and the plant culture in each research. The proposed paper also explains the algorithm solution. Although image processing techniques could identify the diseases, it will classify and quantify as well. The study has been over the last 15 years. Some methods lead to this discussion.

METHODS ARE TOO PRECISE

The common techniques have been able to understand the various rice diseases. A lot of processes had been suggested and were not able to specify only one of the particular diseases in the plant, but those plants need to be explained clearly with at least one of the diseases. Algorithms proposed by several authors are not much effective and many of the papers cannot move to other stages. The algorithms explained are not clear to the specification.

OPERATIONS QUALITY IS TOO EXACT

A lot of the images used to expand the new techniques are collected under very exact conditions, as a regular operation, digital camera capture is used in the early stage of research. This is applicable in real-world applications.

LACK OF SCIENTIFIC RECOGNITION

The simplest solution is explained in image processing techniques. But techniques like SVM, SOM, remote sensing, NVDI, neural networks and fuzzy logic need to more effective: in this case itis not effective.

CONCLUSION

The broad areas of digital images based on image processing techniques are shown in the discussion, but the difficult issues are missed in these cases. Due to the large numbers of citations the explanations are brief and provide a rapid review of the work for the solutions. Some papers added in the references are not included to reduce the length of the chapter. Thus, if the reader needs more information he/she should refer to the bibliography.

REFERENCES

Anthonys, G. & Wickramarachchi, N. (2009). An image recognition system for crop disease identification of paddy fields in Sri Lanka. In *International Conference on Industrial and Information Systems*, Peradeniya, Sri Lanka, pp. 403–407.

Asfarian, A., Herdiyeni, Y., Rauf, A. & Mutaqin, K. H. (2013). Paddy diseases identification with texture analysis using fractal descriptors based on Fourier spectrum. In *International Conference on Computer, Control, Informatics and Its Applications*, Jakarta, Indonesia, pp. 77–81.

Balleda, K., Satyanvesh, D., Sampath, N. V. S. S. P., Varma, K. T. N. & Baruah, P. K. (2014). Agpest: An efficient rule-based expert system to prevent pest diseases of rice & wheat crops. In *8th IEEE International Conference on Intelligent Systems and Control*, Coimbatore, India, pp. 262–268.

Devi, D. A. & Muthukannan, K. (2014). Analysis of segmentation scheme for diseased rice leaves. In *IEEE International Conference on Advanced Communications Control and Computing Technologies*, Ramanathapuram, India, pp. 1374–1378.

Eizenga, G. C., Prasad, B., Jackson, A. K. & Jia, M. H. (2013). Identification of rice sheath blight and blast quantitative trait loci in two different O. sativa/O. nivara advanced backcross populations. *Molecular Breeding*, *31*(4), 889–907.

Huang, J., Wang, H., Dai, Q. & Han, D. (2014). Analysis of NDVI data for crop identification and yield estimation. *IEEE Journal of Selected Topics in Applied Earth Observations and Remote Sensing*, *7*(11), 4374–4384.

Imam, J., Alam, S., Mandal, N. P., Variar, M. & Shukla, P. (2014). Molecular screening for identification of blast resistance genes in North East and Eastern Indian rice germplasm (Oryzasativa L.) with PCR based makers. *Euphytica*, *196*(2), 199—211.

Jingbo, X., Xuehai, H., Feng, S., Xiaohui, N. & Silan, Z. (2009).Prediction of disease-resistant gene by using artificial neural network. In *International Conference on Research Challenges in Computer Science*, Shanghai, China, pp. 81–84.

Joshi, A. A. & Jadhav, B. D. (2016). Monitoring and controlling rice diseases using Image processing techniques. In *International Conference on Computing, Analytics and Security Trends (CAST)*, Pune, pp. 471–476.

Kalita, H., Sarma, S. K. & Choudhury, R. D. (2016). Expert system for diagnosis of diseases of rice plants: Prototype design and implementation. In *International Conference on Automatic Control and Dynamic Optimization Techniques*, Pune, India, pp. 723–730.

Kun, L. & Zhiqiang, W. (2012). Rice Blast Prediction based on gray ant colony and RBF neural network combination model. In *5th International Symposium on Computational Intelligence and Design (ISCID)*, Hangzhou, China, pp. 144–147.

Kurniawati, N. N., Abdullah, S. N. H. S., Abdullah, S. & Abdullah, S. (2009a). Investigation on Image processing techniques for diagnosing paddy diseases. In *International Conference of Soft Computing and Pattern Recognition*, Malacca, Malaysia, pp. 272–277.

Kurniawati, N. N., Abdullah, S. N. H. S., Abdullah, S. & Abdullah, S. (2009b). Texture analysis for diagnosing paddy disease. In *International Conference on Electrical Engineering and Informatics*, Selangor, Malaysia, pp. 23–27.

Liu, Y., Qi, X., Young, N. D., Olsen, K. M., Caicedo, A. L. & Jia, Y. (2015). Characterization of resistance genes to rice blast fungus Magnaportheoryzae in a Green Revolution rice variety. *Journal of Molecular Breeding, 35*(1), 1–8.

Liu, L. & Zhou, G. (2009). Extraction of the rice leaf disease image based on BP neural network. In *International Conference on Computational Intelligence and Software Engineering*, Wuhan, China, pp. 1–3.

Mahesh, H. B., Shirke, M. D., Singh, S., Rajamani, A., Hittalmani, S., … Gowda, M. (2016). Indica rice genome assembly, annotation and mining of blast disease resistance genes. *BMC Genomics, 17*(1), 1–12.

Majid, K., Herdiyeni, Y. & Rauf, A. (2013). I-PEDIA: Mobile application for paddy disease identification using fuzzy entropy and probabilistic neural network. In *International Conference on Advanced Computer Science and information Systems*, Bali, Indonesia, pp. 403–406.

Mercy Nesa Rani, P., Rajesh, T. & Saravanan, R. (2013). Development of expert system to diagnose rice diseases in Meghalaya state. In *5th International Conference on Advanced Computing*, Chennai, India, pp. 8–14.

Mutalib, S., Abdullah, M. H., Abdul-Rahman, S. & Aziz, Z. A. (2016).A brief study on paddy applications with image processing and proposed architecture. In *IEEE Conference on Systems Process and Control (ICSPC)*, Melaka, Malaysia, pp. 124–129.

Orillo, J. W., Dela Cruz, J., Agapito, L., Satimbre, P. J. & Valenzuela, I. (2014).Identification of diseases in rice plant (Oryzasativa) using back propagation artificial neural network. In International Conference on *Humanoid,* Nanotechnology, *Information Technology,* Communication *and* Control, Environment *and* Management *(HNICEM)*, Palawan, Philippines, pp. 1–6.

Phadikar, S. & Goswami, J. (2016). Vegetation indices based segmentation for automatic classification of brown spot and blast diseases of rice. In *3rd International Conference on Recent Advances in Information Technology*, Dhanbad, India, pp. 284–289.

Phadikar, S. & Sil, J. (2008).Rice disease identification using pattern recognition techniques. In *11th International Conference on Computer and Information Technology*, Khulna, Bangladesh, pp. 420–423.

Phadikar, S., Sil, J. & Das, A. K. (2012). Vegetative indices and edge texture based shadow elimination method for rice plant images. In *International Conference on Radar, Communication and Computing*, Tiruvannamalai, India, pp. 1–5.

Prasad, S., Peddoju, S. K. & Ghosh, D. (2014). Energy efficient mobile vision system for plant leaf disease identification. In *IEEE Wireless Communications and Networking Conference (WCNC)*, Istanbul, Turkey, pp. 3314–3319.

Pujari, J. D., Yakkundimath, R. & Byadgi, A. S. (2015). Image processing based detection of fungal diseases in plants. *Procedia Computer Science*, ScienceDirect (Vol. 46, pp. 1802–1808), ISSN 1877-0509.

Qin, Z., Zhang, M., Christensen, T., Li, W. & Tang, H. (2003). Remote sensing analysis of rice disease stresses for farm pest management using wide-band airborne data. In

Proceedings of IEEE Publications International Geoscience and remote Sensing Symposium, Vol. 4, pp. 2215–2217.

Ramkumar, G., Madhav, M. S., Rama Devi, S., Umakanth, B., Pandey, M. K., Prasad, M. S., ... Ravindra Babu, V. (2016). Identification and validation of novel alleles of rice blast resistant gene Pi54, and analysis of their nucleotide diversity in landraces and wild Oryza species. *Euphytica*, *209*(3), 725–737.

Rastogi, A., Arora, R. & Sharma, S. (2015). Leaf disease detection and grading using computer vision technology & fuzzy logic. In *2nd International Conference on Signal processing and integrated Networks*, Noida, India, pp. 500–505.

Ren, Y. et al. (2010). Prediction of disease-resistant gene in rice based on SVM-RFE. In *3rd International Conference on Biomedical Engineering and Informatics*, Yantai, China, pp. 2343–2346.

Robindro, K. & Sarma, S. K. (2013.). JESS based expert system architecture for diagnosis of rice plant diseases: Design and prototype development. In *4th International Conference on Intelligent Systems, Modelling and Simulation*, Bangkok, Thailand, pp. 674–676.

Sanyal, P., Bhattacharya, U., Parui, S. K., Bandyopadhyay, S. K. & Patel, S. (2009). Color texture analysis of rice leaves diagnosing deficiency in the balance of mineral levels towards improvement of crop productivity. In *International Conference on Information Technology (ICIT)*, Orissa, India, pp. 85–90.

Shah, J. P., Prajapati, H. B. & Dabhi, V. K. (2016). A survey on detection and classification of rice plant diseases. In *IEEE International Conference on Current Trends in Advanced Computing*, Bangalore, India, pp. 1–8.

Tanweer, F. A., Rafii, M. Y., Sijam, K., Rahim, H. A., Ahmed, F., Ashkani, S. & Latif, M. A. (2015). Identification of suitable segregating SSR markers for blast resistance in rice using inheritance and disease reaction analysis in backcross families. *Australasian Plant Pathology*, *44*(6), 619–627.

Xiao, W., Yang, Q., Sun, D., Wang, H., Guo, T., Liu, Y., ... Chen, Z. (2015). Identification of three major R genes responsible for broad-spectrum blast resistance in an indica rice accession. *Molecular Breeding*, *35*(1), 1–11.

Yao, Q., Guan, Z., Zhou, Y., Tang, J., Hu, Y. & Yang, B. (2009). Application of support vector machine for detecting rice diseases using shape and color texture features. In *International Conference on Engineering Computation*, Hong Kong, China, pp. 79–83.

Zhou, Y., Wang, Y. & Yao, Q. (2010). Segmentation of rice disease spots based on improved BPNN. In *International Conference on Image Analysis and Signal Processing*, Zhejiang, China, pp. 575–578.

Zhou, Z., Zang, Y., Li, Y., Zhang, Y., Wang, P., & Xiwen Luo (2013). Rice plant-hopper infestation detection and classification algorithms based on fractal dimension values and fuzzy C-means. *Mathematical and Computer Modeling*, *58*(3–4), 701–709.

10 A Multimodal Biosignal Compression Technique for Monitoring Health in Wearable Devices

Ithayarani Pannerselvam

CONTENTS

INTRODUCTION

Some of the most common chronic diseases including sleep disorders [1], mood disorders [2] and epilepsy [3] require multiple biosignals such as electrocardiogram (ECG), electromyogram (EMG) or electroencephalogram (EEG) for diagnosis and treatment. And long-term recordings of these biosignals are required to prevent such chronic disease patients from sudden death [4] by predicting the symptoms earlier. The e-health Internet of Things (IoT) scenario promotes long-term monitoring using micro-sensors and wearable micro-hubs without affecting patients' daily activities. This sensor-based wearable technology provides a sophisticated environment to the patients and helps them to get treatment at home. These micro-sensor nodes are implanted or mounted on the human body to collect the particular biosignal and forward them to a wearable micro-hub or sink thereby establishing the body area communication as depicted in Figure 10.1.

DOI: 10.1201/9781003097518-10

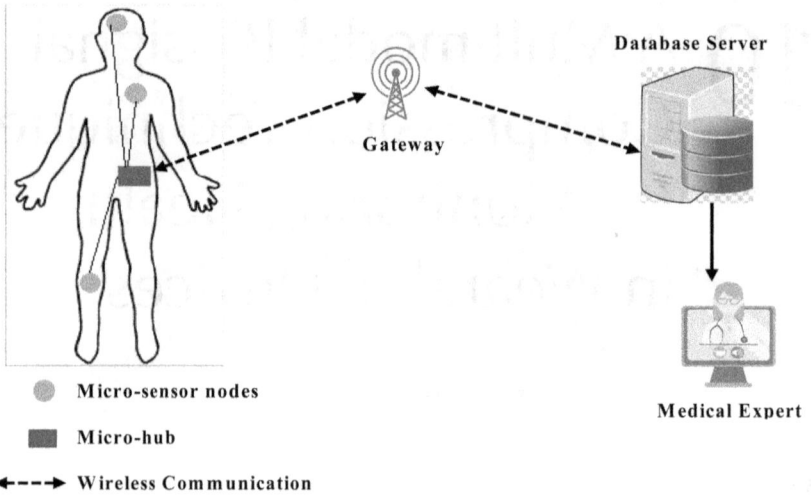

FIGURE 10.1 Schematic diagram of a sensor-based wearable system acquiring multiple biosignals.

Though the sensor nodes can only communicate within a short distance due to the limited transmission range and power [5], the wearable micro-hub, which is enabled with internet technology, collects the readings from the sensor nodes and sends them to the observer. This micro-hub [6] is a resource-constrained battery-relayed device that collects, processes and transmits the signal throughout the day. This prolonged transmission may increase the signal transmission cost and reduces the battery life of wearables. This chapter focuses on compressing the signal before transmission from the sender, i.e., at edge level, to reduce the transmission energy and thereby boosting the battery life of wearables. The objective of this work is to jointly compress the multiple biosignals and reconstruct the novel data from its compressed state with minimal loss. In the case of multiple sensors, compressing signals individually (unimodal compression) may consume more time and energy. This chapter proves that multimodal compression consumes less energy than unimodal compression.

Most of the existing work has been concentrated on the unimodal compression of a biosignal using two emerging paradigms: compressive sensing (CS) [7] and autoencoders (AE) [8]. Recently, some researchers have concentrated on multimodal compression of multiple biosignals using CS algorithms and AE. CS is an efficient algorithm to jointly compress, but it is computationally intensive which is not suitable for lightweight devices and yields a low compression ratio (CR) which cannot effectively reduce the transmission cost. Nevertheless, many of the past works [9, 10] have shown successful results with CS. The algorithm discussed in [11] is a promising technique to jointly compress the multiple biosignals. But these CS and SVD-ASCII algorithms require pre-processing of the biosignal to remove noise as these biosignals are always prone to be contaminated with noise. The concepts discussed in [12] have achieved a good CR of 89% with better reconstruction quality, but the wearables are memory-constrained devices and they could not meet the growing dictionary size when it meets with new signals. However, the algorithms discussed above are good at compression and reconstruction quality, but they lack in focusing on lightweight algorithms for these resource-constrained devices.

A denoising autoencoder (DAE) has been proved to be an efficient algorithm [13] to compress data with high CR at the least computational cost and thereby boosting the battery life of the wearables. As DAE learns deep representation of the data and compresses with a high CR, a multimodal stacked denoising autoencoder (SDAE) [14] is proposed for joint compression and it is most suitable for lightweight wearable devices, because the deep neural network is trained offline and not in the wearable devices. And when it accomplishes decent performance, only the optimal configurations [15] called weights and biases are applied to the device for data processing. The trained network will not grow as in the case of a codebook compression scheme when it meets new incoming data. Previously learned optimised configuration is enough to reconstruct the unknown data and it has been proven experimentally. Moreover, DAE has the ability to denoise signals at the same time, recreating the data as of its compressed form, so DAE does not require any pre-processing step.

However, in the case of deep neural networks the weights are randomly initialised and optimised towards the target value while training, which may increase the training time and may result in poor performance. Transfer learning (TL) [16] may address this issue, it learns the weights of the samples in the source domain and transfers the learned parameters to train the target domains. So, combining TL with deep neural networks may reduce the training time and improve the performance of the system as it uses the previously learned weights. And most researchers have succeeded with a convolutional autoencoder (CAE) for unimodal compression and it is promising for signal reconstruction [17, 18]. So, the TL-based multimodal convolutional denoising autoencoder (M-CDAE) is proposed in this work. Finally, it has been experimentally proven that to achieve better reconstruction quality, a TL-based M-CDAE model is better than standalone (SA) based (M-CDAE) model, i.e., without TL. Also, it is clearly shown that it is better to do multimodal compression than unimodal compression in cases of consuming multiple sensors to utilise the energy efficiently. This chapter is ordered as follows. The basic framework of TL and M-CDAE is presented in the next section. The proposed methodology, M-CDAE with TL, is described in the following section. The experimental results obtained from an appropriate database are then presented, and a comparison of this experiment is conducted. Following this is the conclusion of the chapter.

METHODOLOGY

AE is an ideal methodology for non-linear dimensionality reduction, which typically trains and tests the model with datasets from same scenario, so that the knowledge learned by the model will be limited to that same range of distribution. As this work is a real-time scenario the data range may vary from patient to patient. Therefore, it is necessary to learn different distributions of data which may increase the dataset size for learning. This may result in time-consuming training for the model, also the randomly initialised weights while training the data may yield poor performance. TL can help the model to learn different domains with the knowledge of the previously learned source domain to the target domain. Recent studies made on TL-based AE have achieved outstanding performance by transferring knowledge, in learning a low-dimensional representation for maintenance prediction in intelligent transportation system [19], learning a set of hierarchical non-linear transformations for cross-domain visual recognition [16], real-time production state prediction in the manufacturing industry [20]. Hence, this is the first study based on TL with CDAE for joint compression of multiple biosignals and to effectively reconstruct the signal from its compressed form.

Transfer Learning (TL)

TL overcomes the problem of random weight initialisation while training AE, instead, the training will be initialised by previously learned weights from the source domain. This may train the model for a huge dataset without storing the source data [21], and it promotes adaptation of a trained model for data across various ranges of distribution. In general, TL is characterised into two types, namely instance-based and feature-based [16]. In the case of instance-based only the weights are learned and moved from the source domain to the target domain. And in the case of feature-based the data learned from the source domain is transferred to the target domain. Feature-based transformation is mostly appropriate for classification purposes [22, 23]. For data reconstruction it is better to apply instance-based learning, i.e., only weight transfer. By doing so, the model can be quickly optimised towards the target value which consumes less training time. The structure of AE with TL is given in Figure 10.2. Three steps are involved in instance-based TL, these are pre-training source data, parameters transferring, fine-tuning target data. These three phases are labelled as follows:

1) Pre-training source data: in this phase, AE is trained for the source dataset to determine the optimal weight W_n^{ts} and bias b_n^{ts}. Where n is a number of layers in the AE model, W_n^{ts} and b_n^{ts} are the weight and bias trained for the source dataset.
2) Parameter transferring: following the pre-training step, the learned parameters weight and bias W_n^{ts} and b_n^{ts} are transferred to the new AE model as the initial weight.
3) Fine-tuning model: the transferred weight W_n^{tt} and bias b_n^{tt} are used as an initial weight for the new AE model or for fine-tuning the model. Where W_n^{tt} and b_n^{tt} are the weight and bias trained for target dataset.

Multimodal Autoencoder (M-AE)

The objective of this work is to observe a long-term recording of multiple biosignals (ECG, EMG, EEG) acquired through multiple biosensors mounted on the human

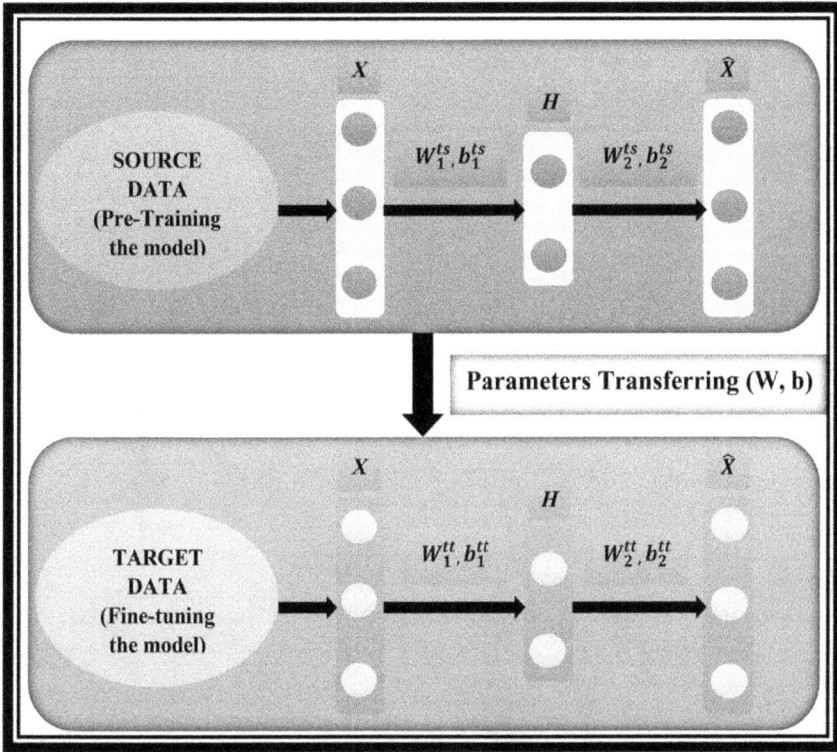

FIGURE 10.2 Architecture of TL-based AE.

body. Figure 10.3 depicts the architecture of deep M-AE. It uses different pathways to jointly compress the multiple inputs to low a dimensional vector called single shared representations. This shared representation will be passed to the decoder (observer or physician) in order to reconstruct the original signal from its compressed form. Also, this single shared representation will be propagated to higher level features independently to reconstruct the original clean input. Once the optimal parameters are obtained through training, then it is configured [15] with a wearable micro-hub for effective delivery of data. This multimodal compression is more efficient than unimodal compression in terms of computation cost, because deep M-AE simultaneously compresses the multiple signals into a single shared representation before transmission which is processed separately in unimodal AE.

The deep M-AE is illustrated in Figure 10.3 which obtains the multiple signals as input and compresses them to a single shared representation and then it is decompressed to original signal through independent pathways. The joint shared representation obtained by the encoder is as follows:

$$h = \sum_{i \in \{c,m,e\}} \tanh\left(x_i^{l-1} * W_i^l + b_i^l\right) \qquad (10.1)$$

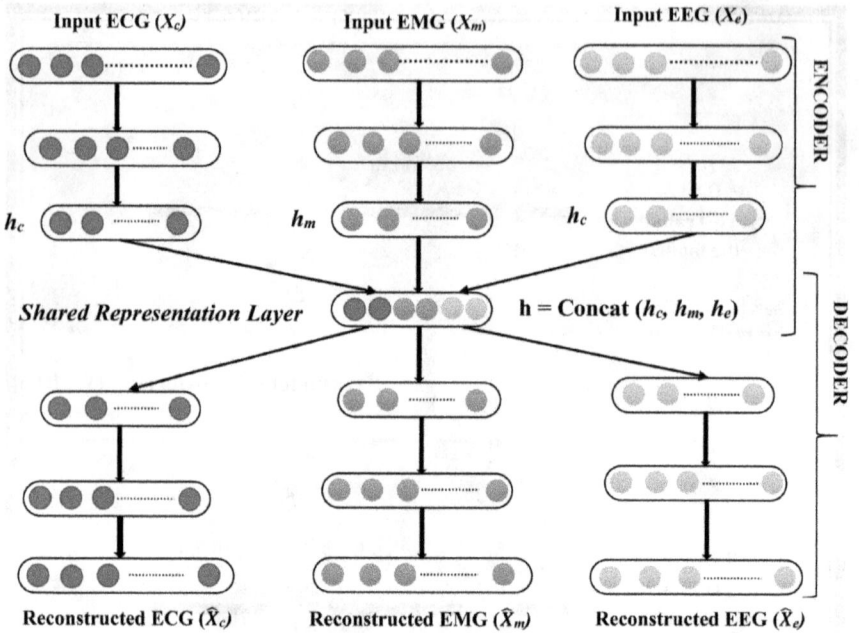

FIGURE 10.3 Schematic representation of deep M-AE.

where c, m, e represents input ECG, EMG, EEG data respectively and l represents the number of layers. x_i^{l-1} is the input from previous layer l. The decoded signals $\hat{x}$ are represented as follows:

$$\hat{x} = \left\{ \hat{x}_{i\in c}^{l} = \tanh\left(h_i^{l-1} * \hat{W}_i^{l} + \hat{b}_i^{l}\right),\ \hat{x}_{i\in m}^{l} = \tanh\left(h_i^{l-1} * \hat{W}_i^{l} + \hat{b}_i^{l}\right), \right.$$

$$\left. \hat{x}_{i\in e}^{l} = \tanh\left(h_i^{l-1} * \hat{W}_i^{l} + \hat{b}_i^{l}\right)\right\} \tag{10.2}$$

Where $\hat{x}$ consists of a set of multiple reconstructed signals and h_i^{l-1} represents the input from previously hidden layer l, where W_i^{l} denotes the weight for current layer l and b_i^{l} represents a bias for current layer l. And tanh ($\bullet$) is the hyperbolic tangent $\left(\dfrac{2}{1+e^{-2z}} - 1\right)$ activation function applied for both encoder and decoder sections.

PROPOSED METHODOLOGY

This work proposes multimodal biosignal compression with TL-based CDAE (M-CDAE). Recent work on TL-based CAE in non-linear dimensionality reduction [25] for image processing in remote sensing, the concept discussed in [22], intelligent urban construction, has been succeeded. In this work the idea behind using TL is to reuse the learned weights from the pre-trained model (source data) for fine-tuning the target data. This may reduce the training and inference cost and increases the signal reconstruction quality which is proven experimentally.

And TL may increase the adaptability of the proposed model for the data across various distributions is proven by reusing the weight which is optimised for different dataset. Also, the comparisons are made between the TL model and SA model. The objective of this work is to efficiently compress the multiple biosignals (ECG, EMG, EEG) and to reconstruct from its compressed form. Moreover, these biosignals are prone to be contaminated with noise due to body movement, so noise removal should be considered. As DAE architecture is employed in the proposed work the signals were simultaneously recovered from its noisy form in the decoder section.

The complete context of the proposed method is depicted schematically in Figure 10.4 and it consists of three pathways for input: ECG, EMG and EEG signals. Before processing the signals, they are additively tarnished by random Gaussian white noise at specified SNR levels of −1, 0, 5 dB. Then they are passed into the model simultaneously, each with the size of 1024 × 1. The proposed model employs CDAE architecture to reduce redundant information and to learn the locally spatial information efficiently by convolutional filters. And maxpool layers which help to reduce the input size, hence the signal of size 1024 × 1 is reduced to the size of 8 × 1 and thereby achieving a CR of 128. The CDAE architecture employed for these three signals will have the same configurations to reveal its multimodal property and the structure of CDAE is depicted clearly in Figure 10.5 and its detailed parameters are given in Table 10.1. Thus, the encoder section efficiently compresses the signals independently but simultaneously. And the encoded data is finally concatenated using a concatenate function before sending them to the decoder section in the representation layer.

The decoder section receives the single shared representation of encoded data of three signals (ECG, EMG, EEG) with sizes of 24 × 1. This shared representation has been fragmented again to three inputs each with sizes of 8 × 1 for decoding, using small anonymous Lambda function. Then the fragmented input has its own pathway for decompression. The compressed input 8 × 1 is reconstructed to 1024 × 1 by an upsampling function which efficiently enhances the data and convolutional filters. The uncorrupted ECG signal is obtained from the high-level features. During the training phase, the output from every dense unit is related with the original uncorrupted version. The gradient loss function optimises the weight and bias towards original clean signals. Thus, the original clean signal is reconstructed independently but simultaneously. This simultaneous process may save more computation time than separate unimodal compressions and it is explained clearly in the experimental results. A hyperbolic tangent function is employed as an activation function throughout the model. And the encoder section is equipped with a batch normalisation (BN) layer and dropout technique with the rate of 0.5 to avoid overfitting.

For training the proposed CDAE model, mean squared error (MSE) is working to update the model's parameter $\theta = \left\{ W, b, \hat{W}, \hat{b} \right\}$ by optimising the following function,

$$L(\theta) = \sum |x - \hat{x}|_2^2 \qquad (10.3)$$

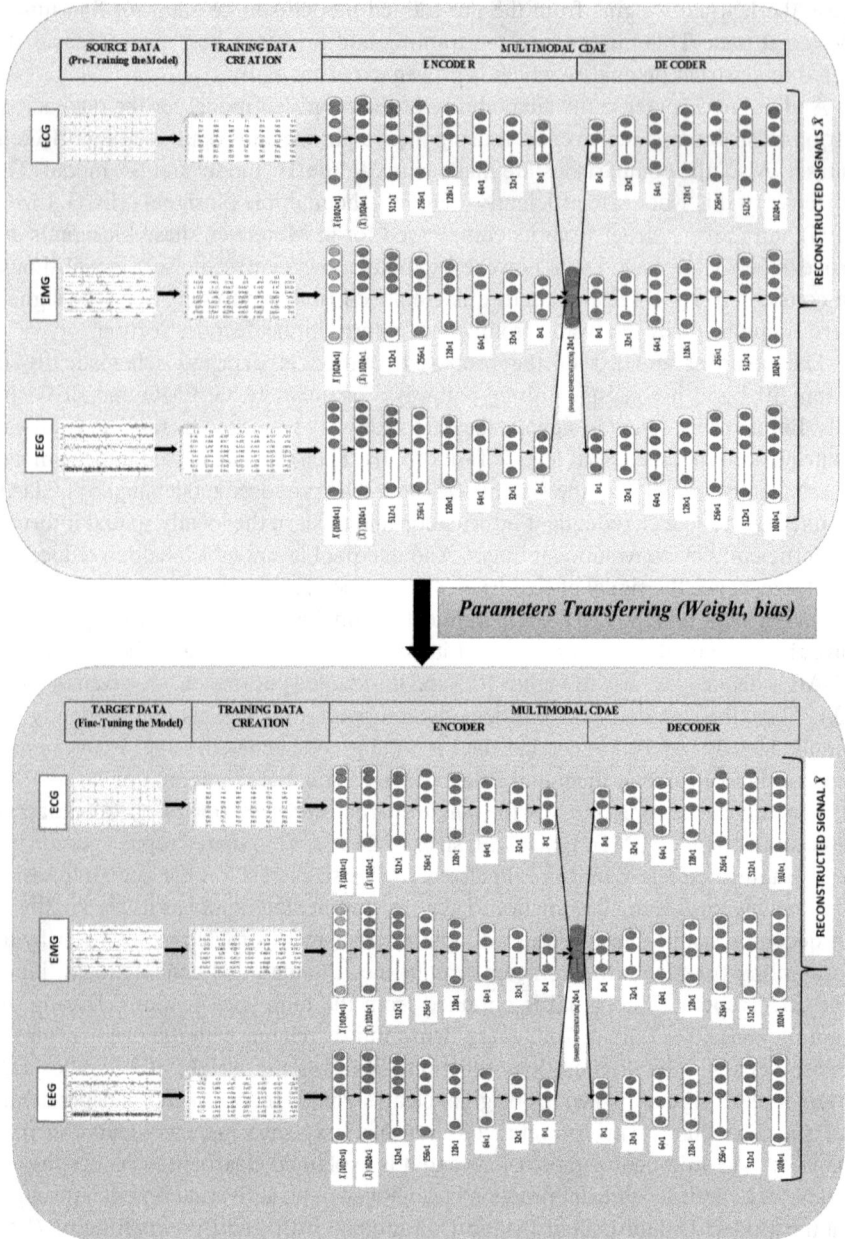

FIGURE 10.4 Framework for the proposed TL-based M-CDAE model.

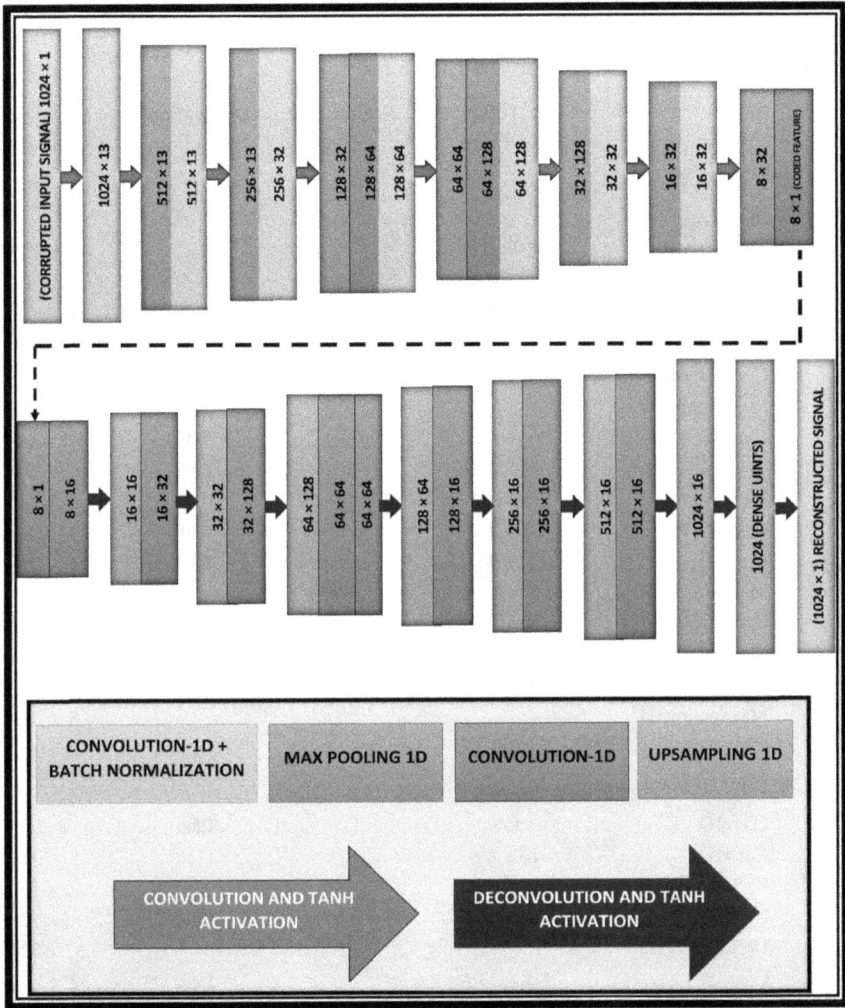

FIGURE 10.5   Detailed structure of the proposed CDAE model.

where W, b are weight and bias in the encoder section and $\hat{W}, \hat{b}$ are weight and bias in the decoder section. Equation (10.3) propagates towards reconstructing the uncorrupted version of the equivalent corrupted input data.

PERFORMANCE ASSESSMENT

PERFORMANCE ASSESSMENT MEASURES

Quality score (QS), compression ratio (CR), root mean square error ($RMSE$), percentage root mean square difference (PRD) and signal to noise ratio (SNR_{imp}) are

TABLE 10.1

Operational Statistics of Proposed CDAE model

S. No.	Layer Name	No. of Filters × Kernel Size	Region/Unit Size	Activation Function	Output Size
ENCODER SECTION					
1	Input Layer	-	-	-	1024×1
2	Conv1D + BN	13×6	-	Tanh	1024×13
3	MaxPooling1D	-	2	-	512×13
4	Conv1D + BN	13×7	-	Tanh	512×13
5	MaxPooling1D	-	2	-	256×13
6	Conv1D + BN	32×7	-	Tanh	256×32
7	MaxPooling1D	-	2	-	128×32
8	Conv1D	64×7	-	Tanh	128×64
9	Conv1D + BN	64×7	-	Tanh	128×64
10	MaxPooling1D	-	2	-	64×64
11	Conv1D	128×13	-	Tanh	64×128
12	Conv1D + BN	128×13	-	Tanh	64×128
13	MaxPooling1D	-	2	-	32×128
14	Conv1D + BN	32×8	-	Tanh	32×32
15	MaxPooling1D	-	2	-	16×32
16	Conv1D + BN	32×8	-	Tanh	16×32
17	MaxPooling1D	-	2	-	8×32
18	Conv1D	1×8	-	Tanh	8×1
DECODER SECTION					
19	Conv1D	1×8	-	Tanh	8 1
20	Conv1D	16×8	-	Tanh	8×16
21	UpSampling1D	-	2	-	16×16
22	Conv1D	32×8	-	Tanh	16×32
23	UpSampling1D	-	2	-	32×32
24	Conv1D	128×13	-	Tanh	32×128
25	Conv1D	128×13	-	Tanh	32×128
26	UpSampling1D	-	2	-	64×128
27	Conv1D	64×7	-	Tanh	64×64
28	Conv1D	64×7	-	Tanh	64×64
29	UpSampling1D	-	2	-	128×64
30	Conv1D	16×8	-	Tanh	128×16
31	UpSampling1D	-	2	-	256×16
32	Conv1D	16×8	-	Tanh	256×16
33	UpSampling1D	-	2	-	512×16
34	Conv1D	16×8	-	Tanh	512×16
35	UpSampling1D	-	2	-	1024×16
36	Dense	-	1024	Tanh	1024×1

ideal criteria for performance evaluation for compression techniques and it is defined as follows:

The algorithm's compression efficiency and its reconstruction quality are determined by QS as given by Equation (10.4).

$$QS = \frac{CR}{PRD} \tag{10.4}$$

The CR ratio will be high for the actual algorithm calculated by finding the ratio amid the size of original signal s_o and the size of compressed signal s_c.

$$CR = \frac{s_o}{s_c} \tag{10.5}$$

The variance amongst the original signal x_i and the restored signal $\hat{x}_i$ is determined by $RMSE$. N is the entire length of the signal. The lower the value of $RMSE$ signifies good performance and smaller variations:

$$RMSE = \sqrt{\frac{1}{N} \times \sum_{n=1}^{N} (x_i - \hat{x}_i)^2} \tag{10.6}$$

PRD is given by Equation (10.7). A lower the PRD value signifies the better quality of the reconstructed signal.

$$PRD = \sqrt{\frac{\sum_{n=1}^{N} (x_i - \hat{x}_i)^2}{\sum_{n=1}^{N} x_i^2}} \times 100 \tag{10.7}$$

SNR_{imp} is applied to observe how well the signal is denoised, it is specified by the following Equation (10.8):

$$SNR_{imp} = SNR_{out} - SNR_{in} \tag{10.8}$$

Where SNR_{in} and SNR_{out} is formulated as follows, here $\tilde{x}_i$ is a corrupted input signal.

$$SNR_{in} = 10 \times \log_{10} \left(\frac{\sum_{n=1}^{N} (x_i)^2}{\sum_{n=1}^{N} (\tilde{x}_i - x_i)^2} \right) \tag{10.9}$$

$$SNR_{out} = 10 \times \log_{10} \left(\frac{\sum_{n=1}^{N} (x_i)^2}{\sum_{n=1}^{N} (\hat{x}_i - x_i)^2} \right) \tag{10.10}$$

EXPERIMENTAL DATA

This work implements three biosignals, namely ECG, EMG and EEG, which have been taken from standard databases and are explained as follows. ECG signals used in this study are derived from the MIT arrhythmia database [25]. The MIT-BIH Arrhythmia Database contains 48 half-hour excerpts of two-channel (lead I and lead II) ambulatory ECG recordings, obtained from 47 subjects. The EEG data used in this study is taken from the Bonn dataset [26] which is sampled at 173.6 Hz with 12-bit resolution. And it is composed of five sets of data which are denoted by Z, O, N, F and S, and each contains 100 single-channel EEG segments. EMG signal recordings, which were recorded at 50 kHz and then downsampled to 4 kHz and exhibited sparsity in both the time and frequency domains, were collected from the Physiobank [27].

The idea behind the work is to transfer the weight learned in source data (pre-training model) to the target data (fine-tuning model) as initial weight, instead of random weight initialisation. This initial weight transfer may reduce the computation burden and make the model most adaptable for different distributions of data. For experiments, the dataset is divided into two parts: source data and target data. And it is clearly charted in Table 10.2. Different records are taken for source data (pre-training model) and target data (fine-tuning model) with equal data size of 100 fragments and each fragment with 1024 samples. The dataset (ECG, EMG, EEG) is split into 80% for training and 20% for testing correspondingly for both the source and target datasets.

EXPERIMENTAL RESULTS

As shown in Table 10.2, the biosignals ECG, EMG, EEG are taken from a particular dataset and split into training and testing sets. Table 10.3 shows the quantitative

TABLE 10.2

Data Separation for the Proposed TL-Based M-CDAE Model

Biosignal	Database	Data Separation	Record No. (or) Name	No. of Records × No. of Fragments	No. of Samples/ Fragment	Training Size (%)	Testing Size (%)
ECG	MIT Arrhythmia database	Source data	100	(1 × 100)	1024	80	20
		Target data	117, 119	(2 × 50)	1024	80	20
EMG	Bonn dataset	Source data	Myopathy	(1 × 100)	1024	80	20
		Target data	Healthy, neuropathy	(2 × 50)	1024	80	20
EEG	Physiobank	Source data	F	(1 × 100)	1024	80	20
		Target data	(Z, O, N, S)	(4 × 25)	1024	80	20

TABLE 10.3

TL-Based M-CDAE vs SA-Based M-CDAE Model with CR of 128

Model Type	Input SNR (dB)	Biosignals	Evaluation Criteria			
			RMSE	PRD	SNR_{imp}	QS
SA Model	−1	ECG	0.0122	0.0575	25	2226
		EMG	0.1154	0.584	4.54	219
		EEG	0.167	1.03	4.7	124
	0	ECG	0.0126	0.06	24.7	2133
		EMG	0.234	1.1	2.1	116
		EEG	0.22	1.04	2.9	123
	5	ECG	0.013	0.0672	20.2	1905
		EMG	0.244	1.1	1.1	116
		EEG	0.159	1.4	1.4	91.4
TL Model	−1	ECG	0.0119	0.05	26.3	2560
		EMG	0.0830	0.42	7.4	305
		EEG	0.153	1	5	128
	0	ECG	0.012	0.055	25.12	2327
		EMG	0.2	1	4.2	128
		EEG	0.158	1.01	4	127
	5	ECG	0.0125	0.063	25	2031
		EMG	0.23	1	2.14	128
		EEG	0.16	1.1	3.89	116

analysis of the test scores of RMSE, PRD and SNR_{imp} for the SA-based M-CDAE model and the TL-based M-CDAE model over the target dataset at specified input SNR noise level of (−1, 0, 5) dB. From the readings, it is observed that the TL-based M-CDAE model yields better results than the SA-based M-CDAE model for all the input SNR levels (−1, 0, 5) dB. Also, it is observed that as the input SNR level increases, the performance of the system degrades. The increased SNR levels increase the RMSE and PRD value and decreases the SNR_{imp} value.

Figure 10.6(a–c) shows the test QS for the SL-based M-CDAE model and proposed TL-based M-CDAE model over the target dataset at specified input SNR noise level of (−1, 0, 5) dB. For the SL-based M-CDAE model, the QS of ECG is 2226 when SNR is −1 dB and it is decreased to 1905 when the SNR level is 5 dB. Similarly, for EMG and EEG when SNR is −1 dB the QS is 219 and 124 respectively and when the SNR level increased to 5 dB the QS will be 116.4 and 124 respectively. These measurements are slightly improved through the proposed TL-based M-CDAE model but undergoing the same issue, as the SNR level increases the performance degrades. For the TL model, the QS of ECG is 2298 when SNR is −1 dB and it is decreased to 2370 when the SNR level is 5 dB. Similarly, for EMG and EEG when SNR is −1 dB the QS is 305 and 128 respectively and when the SNR level is increased to 5 dB the QS will be 128 and 116 respectively [28].

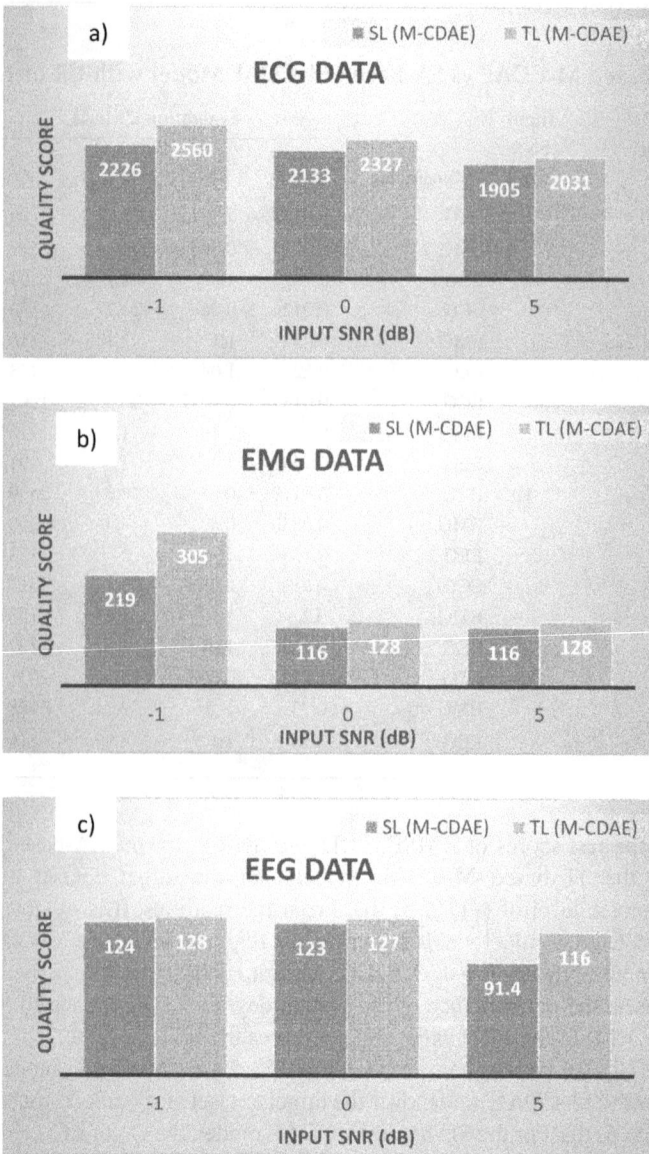

FIGURE 10.6 Comparison of the quality scores for TL-based M-CDAE and SA-based M-CDAE models of all biosignals (ECG, EMG, EEG) at different input SNR levels of (−1, 0, 5) dB. (a) Quality scores of ECG. (b) Quality scores of EMG. (c) Quality scores of EEG [29].

Figure 10.7 represents the overall graphical visualisation of the waveform of ECG, EMG, EEG biosignals for the proposed TL-based M-CDAE model. And Figure 10.8(a–f) clearly shows the comparison results of the clean and reconstructed output of ECG, EMG, EEG biosignals produced by the SA-based M-CDAE model

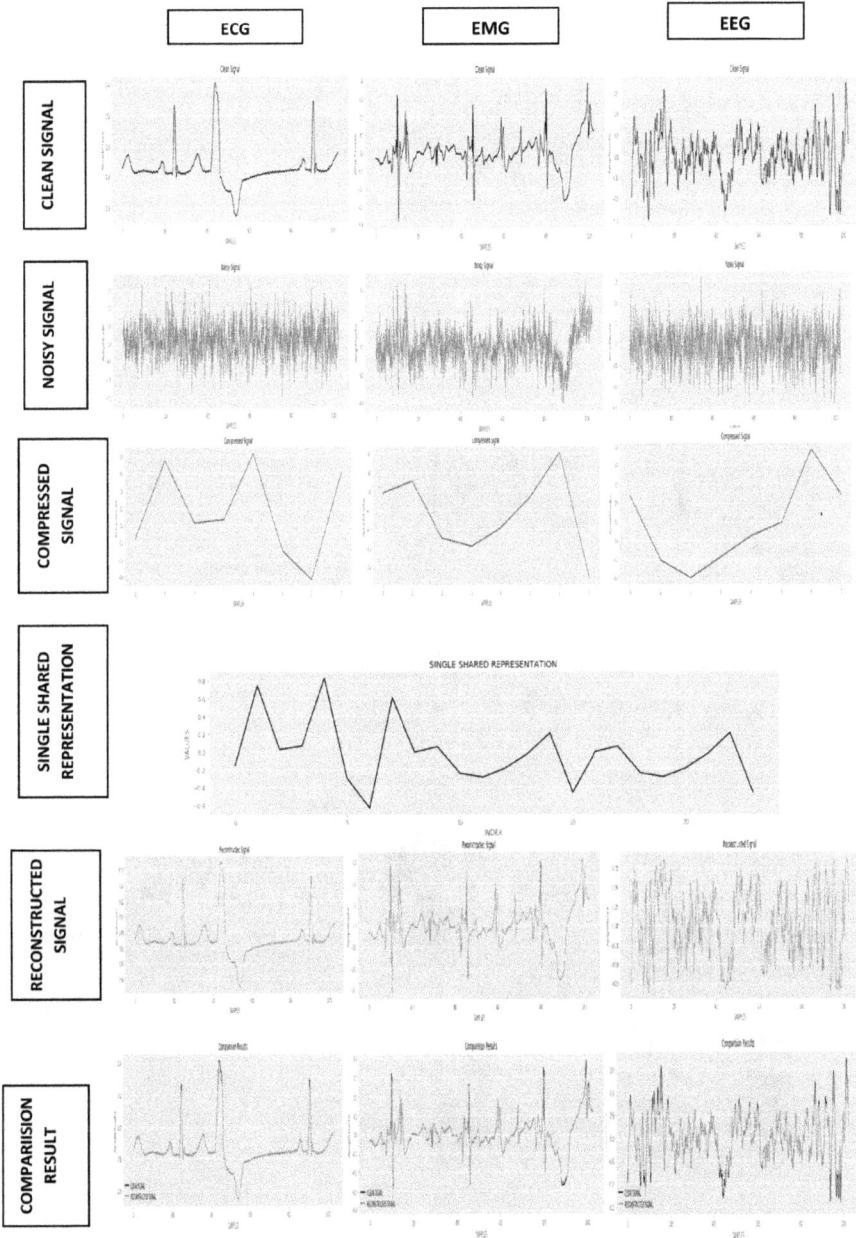

FIGURE 10.7 Overall graphical representation of reconstructed results of ECG, EMG, EEG biosignals for the proposed TL-based M-CDAE model.

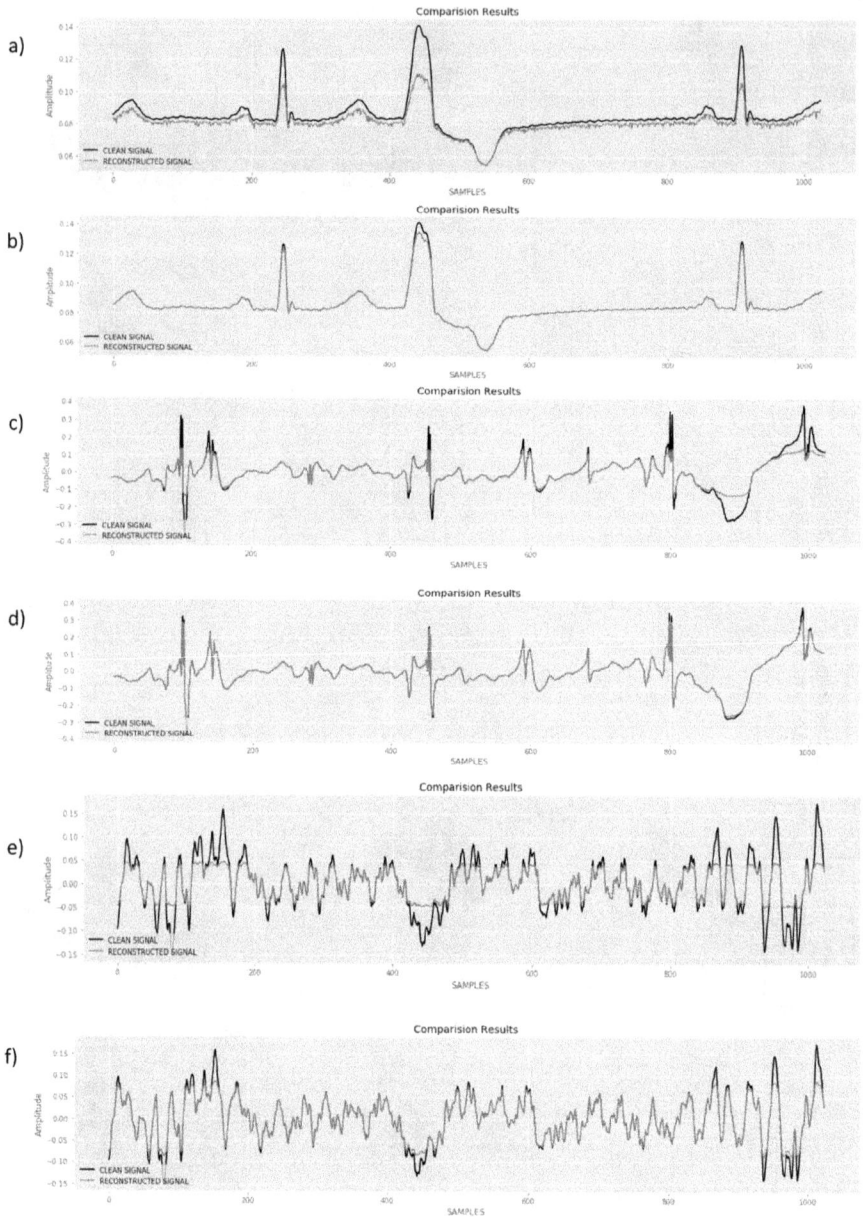

FIGURE 10.8 Evaluation results of clean and reconstructed biosignals over the target dataset with and without TL. (a) ECG signal reconstructed by SA-based M-CDAE model. (b) ECG signal reconstructed by TA-based M-CDAE model. (c) EMG signal reconstructed by SA-based M-CDAE model. (d) EMG signal reconstructed by TA-based M-CDAE model, (e) EEG signal reconstructed by SA-based M-CDAE model. (f) EEG signal reconstructed by TA-based M-CDAE model.

TABLE 10.4

SA-Based CDAE Model vs TL-Based CDAE Model for Unimodal and Multimodal Compression in Terms of Computation Cost at the Input SNR Level of 5 dB

| Compression | | SA CDAE Model | | TL CDAE Model | |
| | | **Computation Time (Seconds)** | | **Computation Time (Seconds)** | |
TYPE	Biosignal	Training Cost	Inference Cost	Training Cost	Inference Cost
Unimodal	ECG	5482.35	1.15	4961.86	0.0418
	EMG	6158.89	1.4	5117.86	0.323
	EEG	3923.38	0.7319	3614.98	0.455
Total		15,564.62	3.28	13,694.7	0.8198
Multimodal	ECG-EMG-EEG	8363.3	0.7	8211.92	0.1007

and the TL-based M-CDAE model over the target dataset as given in Table 10.2. According to the tabulated results and graphical results, the TL-based M-CDAE model outperforms in all input SNR levels of (−1, 0, 5) dB.

Energy efficiency is not a novel issue, but it is the issue that should be considered in wireless communication. Since the wearable hub, which is responsible for acquiring and transmitting the multiple biosignals, is miniaturised and battery-relayed, the runtime of the signal processing algorithm should be considered. This work focuses on jointly compressing the multiple biosignal in earlier transmission to decrease the transmission cost. In the case of acquiring and transmitting multiple biosignals, it is better to do multimodal compression than unimodal compression. Because the training and testing (inference) time of multimodal compression is lower than unimodal compression, as shown in Table 10.4, for a highly corrupted input signal at an SNR level of 5 dB [30].

DISCUSSION

Multimodal compression has been carried out by most of the studies [9–12] but this is the first study to propose TL-based multimodal compression for multiple biosignals ECG, EMG and EEG. And each signal is taken from standard databases. This work performs four types of experiments; with the proposed CDAE model are SA-based unimodal compression, SA-based multimodal compression, TL-based unimodal compression and TL-based multimodal compression. The experimental result proves that TL-based multimodal compression will be better in terms of computation cost and QS. The reason behind the successive result is, in the transfer learning model the learned weight is reused so that the system can soon propagate towards the optimal configurations (weights and bias). Also, while compressing the multiple signals simultaneously the training time and inference cost is reduced further as shown in Table 10.4. Most of the work achieved great success in signal reconstruction using CDAE so this work employs improved CDAE architecture and achieves a CR of 128 [31].

The dataset used for experiments has been separated into source and target datasets and the proposed model is pre-trained on the source dataset. The learned weights and bias from the source dataset are reused for training the target dataset. In neural networks, rather than initialising weights randomly it is better to reuse the learned weights which may reduce the training cost and inference cost. Also, by transferring the parameters like weight and bias the model is trained for large data without storing the source data. The experimental results produced here are the test score over target dataset.

Moreover, these biosignals are always prone to be contaminated with noise due to the body's movement. So, the input signal is additively corrupted by random gaussian white noise at various noise levels (−1, 0, 5) dB. DAE architecture is utilised with the proposed work which efficiently denoises the signal at the same time while recreating from its compressed form. Hence, the proposed work can simultaneously denoise signals while reconstructing them from its compressed form.

CONCLUSION

Wearable technology enriches the long-term monitoring system through wearables and wireless communication. This wireless communication always suffers from signal transmission cost due to acquiring and sending multiple signals throughout the day. So, the signals can be compressed before sending at the edge level, i.e., where they are acquired. The wearable hub which is responsible for acquiring and transmitting signals is miniaturised in nature and resource-constrained. So, the designed algorithm should be lightweight and at the same time it should be efficient to reconstruct the signal from its compressed form. This chapter employs TL-based M-CDAE to compress the signal efficiently. In neural networks, rather than initialising the weight randomly it is better to reuse the learned weights so that the model performs well for data across various ranges of distribution. The investigational result depicts that the proposed model consumes less in training and inference cost when performing multimodal compression and gives better QS with a CR of 128 with the TL-based model. Moreover, the proposed system reveals its multimodality property by employing the same CDAE architecture for reconstructing three biosignals, ECG, EMG and EEG. As this work focuses on real-time application, future work can be carried out for on-device compression so that the system can perform well for any new incoming unknown data without offline training.

REFERENCES

1. Chambon, S., Galtier, M. N., Arnal, P. J., Wainrib, G. & Gramfort, A. (2018). A deep learning architecture for temporal sleep stage classification using multivariate and multimodal time series. *IEEE Transactions on Neural Systems and Rehabilitation Engineering*, 26, 758–769. doi: 10.1109/TNSRE.2018.2813138.

2. Jin, L., Zhang, Y., Wang, X.-L., Zhang, W.-J., Liu, Y.-H. & Jiang, Z. (2017). Postictal apnea as an important mechanism for SUDEP: A near-SUDEP with continuous EEG-ECG-EMG recording. *Journal of Clinical Neuroscience, 43*, 130–132. doi: 10.1016/j.jocn.2017.04.035.

3. Klosch, G., Kemp, B., Penzel, T.,Schlogl, A., Rappelsberger, P., Trenker, E., … Dorffner, G. (2001). The SIESTA project polygraphic and clinical database. *IEEE Engineering in Medicine and Biology, 20*, 51–57. doi: 10.1109/51.932725.

4. Tomasini, M., Benatti, S., Milosevic, B., Farella, E. & Benini, L. (2015). Power Line Interference Removal for High Quality Continuous Bio-Signal Monitoring with low-power wearable devices. *IEEE Sensors Journal, 16*, 3887–3895. doi: 10.1109/JSEN.2016.2536363.

5. Xinmeng, X. U., Ning, Z., Houbing, S., Anfeng, L., Ming, Z. & Zeng, Z. (2018). Adaptive beaconing based MAC protocol for sensor based wearable system. *IEEE Access, 6*, 29700–29714. doi: 10.1109/ACCESS.2018.2843762.

6. Samanta, A., Samaresh, B. & Misra, S. (2015). Link quality-aware resource allocation with LoadBalance in wireless body area networks. *IEEE Systems Journal, 12*, 74–81. doi: 10.1109/JSYST.2015.2458586.

7. Craven, D., McGinley, B., Kilmartin, L., Glavin, M. & Jones, E. (2014). Compressed sensing for bioelectric signals: A review. *IEEE Journal of Biomedical and Health Informatics, 19*, 529–540. doi: 10.1109/JBHI.2014.2327194.

8. Del Testa, D. & Rossi, M. (2015). Lightweight lossy compression of biometric patterns via denoising autoencoders. *IEEE Signal Processing Letters, 22*, 2304–2308. doi: 10.1109/LSP.2015.2476667.

9. Singh, A. & Dandapat, S. (2017). Block sparsity-based joint compressed sensing recovery of multi-channel ECG signals. *Healthcare Technology Letters, 4*, 50–56. doi: 10.1049/htl.2016.0049.

10. Dixon, A. M. R., Allstot, E. G., Gangopadhyay, D. & Allstot, D. J. (2012). Compressed sensing system considerations for ECG and EMG wireless biosensors. *IEEE Transactions on Biomedical Circuits and Systems, 6*, 156–166. doi: 10.1109/TBCAS.2012.2193668.

11. Mukhopadhyay, S. K., Omair Ahmad, M. & Swamy, M. N. S. (2018). SVD and ASCII character encoding-based compression of multiple biosignals for remote healthcare systems. *IEEE Transactions on Biomedical Circuits and Systems, 12*, 137–150. doi: 10.1109/TBCAS.2017.2760298.

12. Carotti, E. S. G., De Martin, J. C., Farina, R. M. D. & Farina, D. (2009). Compression of multidimensional biomedical signals with spatial and temporal codebook-excited linear prediction. *IEEE Transactions on Bio-Medical Engineering, 56*, 2604–2610. doi: 10.1109/TBME.2009.2027691.

13. Hooshmand, M., Zordan, D., Del Testa, E. G. D., Rossi, M. & Rossi, M. (2017). Boosting the battery life of wearables for health monitoring through the compression of biosignals. *IEEE Internet of Things Journal, 4*, 1647–1662. doi: 10.1109/JIOT.2017.2689164.

14. Cao, Y., Zhang, H., Choi, Y.-B., Wang, H. & Xiao, S. (2020). Hybrid deep learning model assisted data compression and classification for efficient data delivery in mobile health applications. *IEEE Access, 8*, 94757–94766. doi: 10.1109/ACCESS.2020.2995442.

15. Ben Said, A., Mohamed, A., Elfouly, T., Harras, K. & Jane Wang, Z. (2017). Multimodal deep learning approach for joint EEG-EMG data compression and classification. In *IEEE Wireless Communications and Networking Conference (WCNC)*.

16. Begam, S., Selvachandran, G., Ngan, T. T. & Sharma, R. (2020). Similarity measure of lattice ordered multi-fuzzy soft sets based on set theoretic approach and its application in decision making. *Mathematics, 8*(8), 1255.

17. Thanh, V., Rohit, S., Raghvendra, K., Le Hoang, S., Thai, P. B., Dieu, T. B., … Le, T. (2020). Crime rate detection using social media of different crime locations and twitter

part-of-speech tagger with brown clustering. *Journal of Intelligent & Fuzzy Systems*, *38*(4), 4287–4299.

18. Nguyen, P. T., Ha, D. H., Avand, M., Jaafari, A., Nguyen, H. D., Al-Ansari, N., ... Pham, B. T. (2020). Soft computing ensemble models based on logistic regression for groundwater potential mapping. *Applied Sciences*, *10*(7), 2469.

19. Jha, S., Kumar, R., Hoang Son, L., Abdel-Basset, M., Priyadarshini, I., Sharma, R. & Viet Long, H. (2019). Deep learning approach for software maintainability metrics prediction. *IEEE Access*, *7*, 61840–61855.

20. Sharma, R., Kumar, R., Sharma, D. K., Le Hoang, S., Priyadarshini, I. & Pham, B. T. (2019). Dieu Tien Bui & Sakshi Rai. Inferring air pollution from air quality index by different geographical areas: Case study in India. *Air Quality, Atmosphere and Health*, *12*(11), 1347–1357.

21. Sharma, R., Kumar, R., Singh, P. K., Raboaca, M. S.& Felseghi, R.-A. (2020). A systematic study on the analysis of the emission of CO, CO_2 and HC for four-wheelers and its impact on the sustainable ecosystem. *Sustainability*, *12*(17), 6707.

22. Sharma, S., Kumar, R., Das Adhikari, J., Mohapatra, M., Sharma, R., Priyadarshini, I. & Le, D. N. (2020). Global forecasting confirmed and fatal cases of COVID-19 outbreak using autoregressive integrated moving average model. *Public Health*. doi: 10.3389/fpubh.2020.580327.

23. Malik, P. et al. (2021). *Industrial Internet of things and its applications in Industry 4.0: State-of the art, computer communication* (Vol. 166, pp. 125–139), Elsevier.

24. Sharma, R., Kumar, R., Satapathy, S. C., Al-Ansari, N., Singh, K. K., Mahapatra, R. P., ... Pham, B. T. (2020). Analysis of water pollution using different physicochemical parameters: A study of Yamuna River. *Frontiers in Environmental Science*, *8*. doi: 10.3389/fenvs.2020.581591, PubMed: 581591.

25. Dansana, D., Kumar, R., Parida, A., Sharma, R., Adhikari, J. D., Van Le, H., ... Pradhan, B. (2021). Using susceptible-exposed-infectious-recovered model to forecast coronavirus outbreak. *Computers, Materials and Continua*, *67*(2), 1595–1612.

26. Vo, M. T., Vo, A. H., Nguyen, T., Sharma, R. & Le, T. (2021). Dealing with the class imbalance problem in the detection of fake job descriptions. *Computers, Materials and Continua*, *68*(1), 521–535.

27. Sachan, S., Sharma, R. & Sehgal, A. (2021). Energy efficient scheme for better connectivity in sustainable mobile wireless sensor networks. *Sustainable Computing: Informatics and Systems*, *30*, PubMed: 100504, 1–11.

28. Ghanem, S., Kanungo, P., Panda, G., Satapathy, S. C. & Sharma, R. (2021). Lane detection under artificial colored light in tunnels and on highways: An IoT-based framework for smart city infrastructure. *Complex and Intelligent Systems*. doi: 10.1007/s40747-021-00381-2.

29. Sachan, S., Sharma, R. & Sehgal, A. (2021). SINR based energy optimization schemes for 5G vehicular sensor networks. *Wireless Personal Communications*. doi: 10.1007/s11277-021-08561-6.

30. Priyadarshini, I., Mohanty, P., Kumar, R., Sharma, R., Puri, V. & Singh, P. K. (2021). A study on the sentiments and psychology of twitter users during COVID-19 lockdown period. *Multimedia Tools and Applications*. doi: 10.1007/s11042-021-11004-w

31. Azad, C., Bhushan, B., Sharma, R., Shankar, A., Singh, K. K. & Khamparia, A. (2021). Prediction model using SMOTE, genetic algorithm and decision tree (PMSGD) for classification of diabetes mellitus. *Multimedia Systems*. doi: 10.1007/s00530-021-00817-2

11 A Comprehensive Study on Metaheuristics, Big Data and Deep Neural Network Strategies

Sushree Swagatika Jena, Sushree Bibhuprada and B. Priyadarshini

CONTENTS

INTRODUCTION

METAHEURISTICS AND BIG DATA

The word 'heuristic' can be defined as the systematic analytical problem-solving approach that basically implements shortcut techniques and sufficiently good calculations to generate satisfactory presentable solutions where a limited time span is provided. The word '**heuristic**' is descended from a Greek word whose original elucidation is 'to discover'. While dealing with particularly composite aggregated data this approach is used to make fast observant decisions or get well-timed determined results. Now, we will understand the meaning of '**metaheuristic**'. In simple words, meta indicates 'higher level' or 'beyond' and heuristic represents 'to find' or 'to know' [1–3].

In the scope of computer science and mathematical optimisation, metaheuristic is a method or mechanism created and formulated at a higher level to generate an algorithm that will prepare and deliver an adequate and acceptable solution to an optimisation problem exceptionally with incomplete, unrefined and partial information or narrow enumeration and estimation ability. This is a large set of solutions to illustrate. In this process, some assumptions are made related to the optimisation problem that is being solved so as to make the statement valid for a diverse collection of problems. Metaheuristics do not guarantee to provide optimal solutions for a class of problems [4–12].

Then comes the term '**big data**'. which indicates to all the data that is being generated is at an unprecedented rate. This particular term '**big data**' came into use since 1990 and it was popularised by John Mashey. The datasets that have a size beyond the size limits of many software applications and tools are put under this term and the size of the data is constantly growing at a rapid rate. Big data constitutes structured, semi-structured and unstructured data and more emphasis is put on unstructured data. It is very important to convert big data into business intelligence that can be readily established and utilised by the enterprises. It is characterised by volume, variety, velocity and the veracity of the data, and these are represented as the four Vs of big data as shown in Figure 11.1. There are diverse sources of big data, therefore various updated technologies are required to deal with it [13, 2].

Short description of the four Vs of big data:

(a) Volume: it is the enormous amount of data that are collected for various purposes and hence the volume becomes a censorious factor while analysing big data.

(b) Velocity: due to the increase in demand and day-to-day activities, surplus data are generated at an exceptional rate and it is very important to manage and parse the big data.

(c) Variety: the data are generated in various ways and in various formats like videos, photos, texts, numeric, etc. They are completely heterogeneous in nature and thereby provide variety to the big data.

(d) Veracity: this can be defined as accuracy. It just explains how truthful and meaningful the collected data are, so that they can be put into use.

Now we will understand how a metaheuristic approach and big data are related.

In order to move ahead in the process of extraction of meaning and information from the big data, complex analytical techniques are required. Therefore,

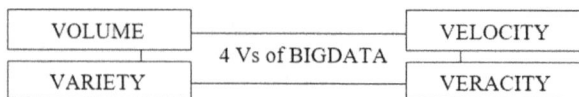

FIGURE 11.1 Four Vs of big data.

metaheuristics are required to find the solution to these challenges, as they are easily adapted to different contexts and can properly execute and succeed in dealing with large sized problems. Metaheuristics can be put into use to develop, enhance and as an upgrading technique to precisely analyse big data in various fields like railway engineering, data mining, medicines etc. Figure 11.2 represents the performance of a metaheuristic approach.

Metaheuristics accumulatively consist of a series of different types of algorithms which include: evolutionary algorithm, trajectory algorithm, naturally inspired algorithm, etc. Some of the algorithms will be described further.

DEEP NEURAL NETWORK (DNN)

A neural network (NN) can be defined as a network consisting of several branches of neurons. In the world of technology this can be properly described as the technology

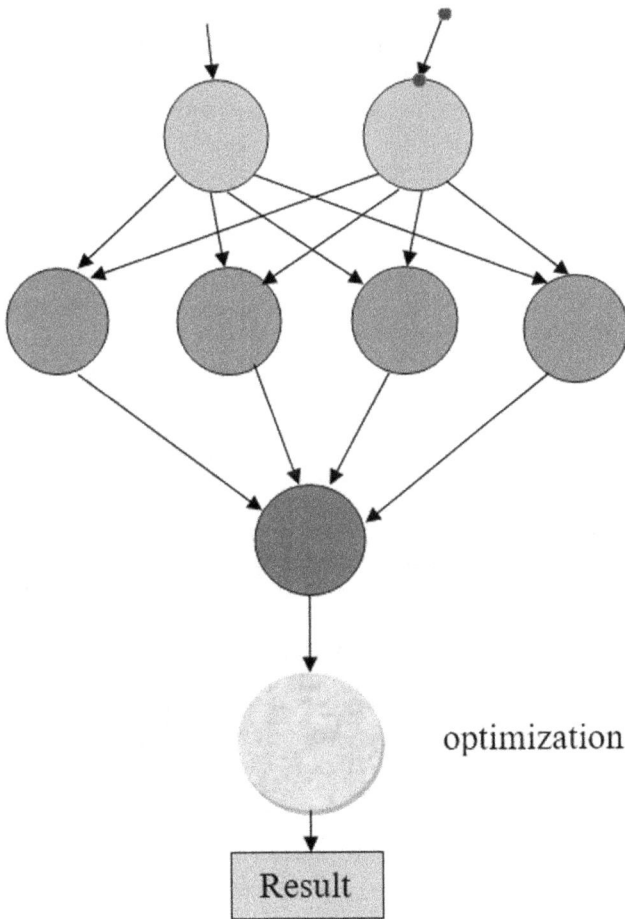

FIGURE 11.2 Metaheuristic approach.

that is developed to replicate and perform the activities of the human brain. A DNN is basically an artificial neural network (ANN). Deep learning (DL) is also employed to describe a DNN. In this case, between the relevant input and output layers there is the presence of numerous layers. The individual layer does particular types of actions that considers sorting jointly with ordering. To attain the desired outcome from the concerned input, the DNN acts to attain the accurate analytical desirable operation. The individual direction and operation is considered as a layer (as stated above) as portrayed in Figure 11.3 [2, 3].

DNN finds the correct output for the given input performing proper mathematical operations in the right directions. This framework produces models where the object exhibits all the compositions of the previous primary layers. So, the extra hidden layers help in storing the features of the lower layers in the upper ones, thereby, enabling modelling of the complex data using lesser units.

A branch of machine learning is also known as DL and is executed and implemented by the DNN. Figure 11.4 shows the relation between machine learning and DL. The word 'deep' has been used as there is the presence of multiple hidden layers. DL has various designs and frameworks which are being utilised in various fields like audio identification, speech realisation and identification, various material analyses and investigations, etc. and have generated very useful results [4–10].

It consists of an unrestricted number of layers with restricted size and it authorises practical execution and optimised performance. When it is provided with raw input, numerous layers are used to abstract and select the high-level characteristics and attributes. Each layer transforms the input into a moderately composite illustration or description. One of the most important facts is that the DL process knows to place the features at the best-suited layer on its own. DL systems consist of credit assignment path (CAP) depth which elucidates the connection between the input and output as it is stated as the series of transformations from input to output. The DL framework can be established with a greedy layer-by-layer strategy.

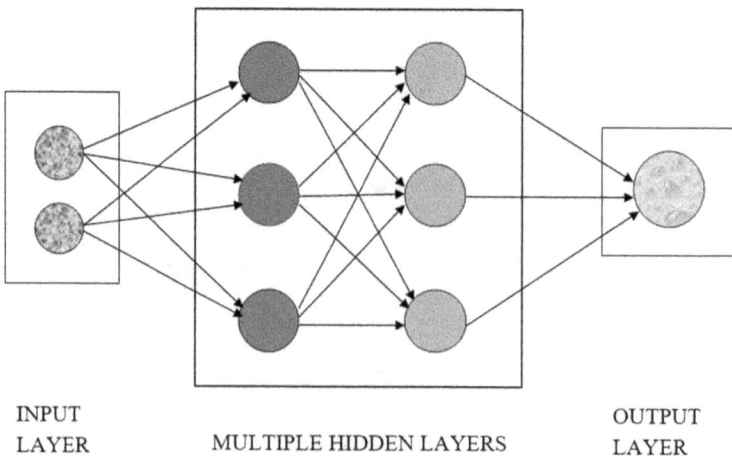

| INPUT LAYER | MULTIPLE HIDDEN LAYERS | OUTPUT LAYER |

FIGURE 11.3 Layered structure of DNN.

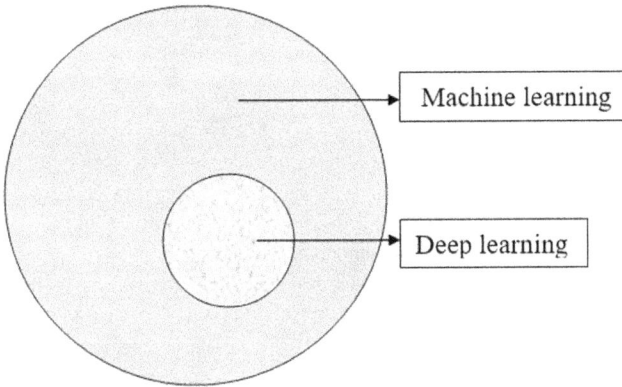

FIGURE 11.4   Relationship between machine learning (ML) and deep learning (DL).

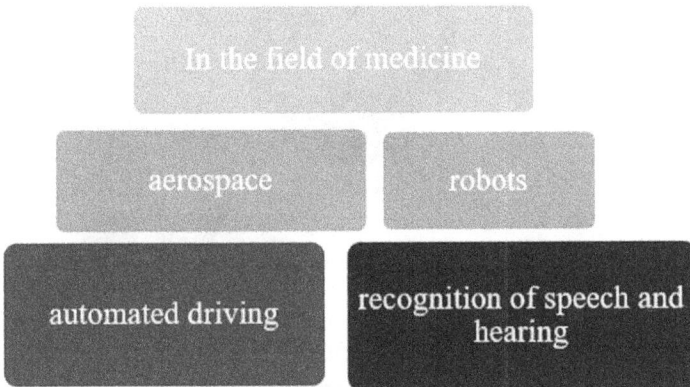

FIGURE 11.5   Usage of DL.

A Brief History of DL

DL is employed in most of the real-world scenarios (refer to the diagram in Figure 11.5). The crucial application areas are as follows:

- DL is applied by cancer analysts for ensnaring the cancer cells automatically in the case of a biomedical scenario.
- DL is employed in automated driving.
- DL is applied for speech recognition in translation and automated hearing.
- DL is employed to track distinct objects from the satellites in aerospace engineering.
- DL is applied for the robotic application that can teach a robot to accomplish a task which is afforded as the input [6, 14–18].

   Work done in the field of DL in various years is illustrated in Table 11.1.

TABLE 11.1
Work Done in DL

Year	Work Done
1967	Alexey Ivakhnenko published a working algorithm for deep, feedforward perceptrons.
1971	A paper illustrating and explaining the eight-layered deep network was published. Kunihiko Fukushima described other framework and structures of DL.
2000	Igor Aizenberg and colleagues introduced DL to ANN.
2009	An assumption was made that a deep neural network can be practically implemented and it was believed that it can be used to overcome the main difficulties of NNs.
2010	Scientists and researchers used DL for large vocabulary recognition and used the large-sized output layers of DNNs.

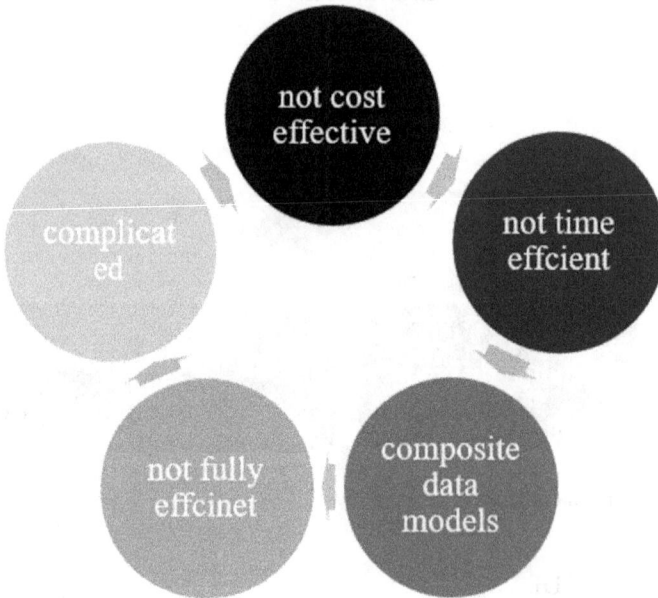

FIGURE 11.6 Disadvantages of DL.

There are some disadvantages of DL as represented in Figure 11.6.
The following are some of the disadvantages of DL:

- It is not very cost effective as it requires a huge amount of memory space and other electrical and computational resources.
- It is complicated, not understood easily and really very hard to narrate.
- It is not fully efficient, and it does not provide us with particular and 100% correct results.
- Large datasets are needed to train due to the composite data models.
- It is not time efficient.

FEED FORWARD NEURAL NETWORK (FNN)

FNN is the ANN in which the nodes are interconnected in such a way that they do not form a cycle or a loop. If in any way a cycle is getting formed as shown in Figure 11.7 then that will not be treated as a FNN [15, 18]. It is the uncomplicated ANN which was designed at the beginning and there is one-directional movement of the information, that is from input to output, passing though the hidden layers [6, 19, 20, 21].

A **directed acyclic graph**, that is comprised of a finite number of vertices and edges, with direction from one vertex to another vertex such that there is no formation of any cycle, can be used as the feed forward network and some vertices will be selected as input and others will be output. A **radial basis function** network can be treated as the FNN. It is basically an ANN, and as the activation function it specifically uses the radial basis functions [22–36].

APPLICATION OF METAHEURISTICS ALGORITHMS

APPLICATION OF METAHEURISTIC ALGORITHM IN NN TRAINING

Previously, we have understood what a metaheuristic algorithm is and how it works. Now we will focus on the uses of a metaheuristic algorithm in various fields. This particular algorithm has been devised and executed in the NN training in an outstanding manner. It is said that training NNs is an important and a very complicated task. In order to speed up the training process of the NN, some metaheuristic optimisation methods have been used and are found to be productive [1–3, 37–42].

One such algorithm is the CK algorithm or cuckoo search algorithm which is newly developed and is appropriate for resolving optimisation problems. Figure 11.8 shows the generalised process of NN training, which will deliver the basic idea about how this technique works. The following are some of the crucial steps involved in this process:

- ANN is booted up or loaded with weights and biases provided by the metaheuristic optimisation algorithm.
- The input from various documents is provided to the network.

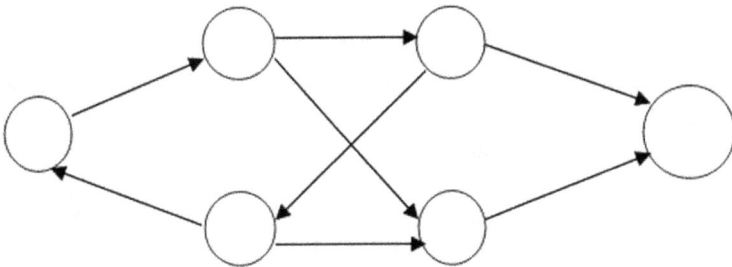

FIGURE 11.7 Network connectivity.

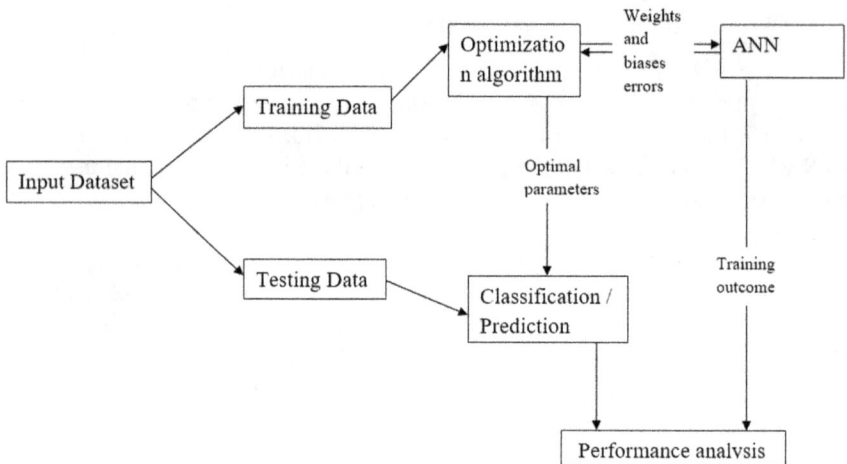

FIGURE 11.8   Generalised process of NN training.

- The output that is produced is matched to the desired output.
- The error value that is obtained is regarded as the fitness value.
- Minimisation of function is done to get minimal fitness value.
- The processes that are mentioned above are carried out until it meets the stopping criteria.
- The answer that has the minimal error or the best fitness value is chosen as the appropriate solution.

GENETIC ALGORITHM (GA) AND NN

GA is a subset of the huge class of evolutionary algorithms. It was established in 1975 by Professor Holland. It is formulated on Darwin's theory of evolution which specifically defines the gene selection and natural elimination procedure. This algorithm is particularly used to produce solutions of good standard and quality to the optimisation and search problems that are based on the various biological processes like chromosome mutation, crossovers and selection operation [43, 44].

Figure 11.9 portrays a genetic algorithm flow diagram.

The steps are described below:

Step 1: chromosomes which contain all the details and information are indiscriminately created.
Step 2: fitness value is generated using all the chromosomes and the fitness functions.
Step 3: for the upcoming generation, chromosomes are selected as parents in a random manner without following any specific rule.
Step 4: crossover process is carried out by the parents to obtain the next generation.

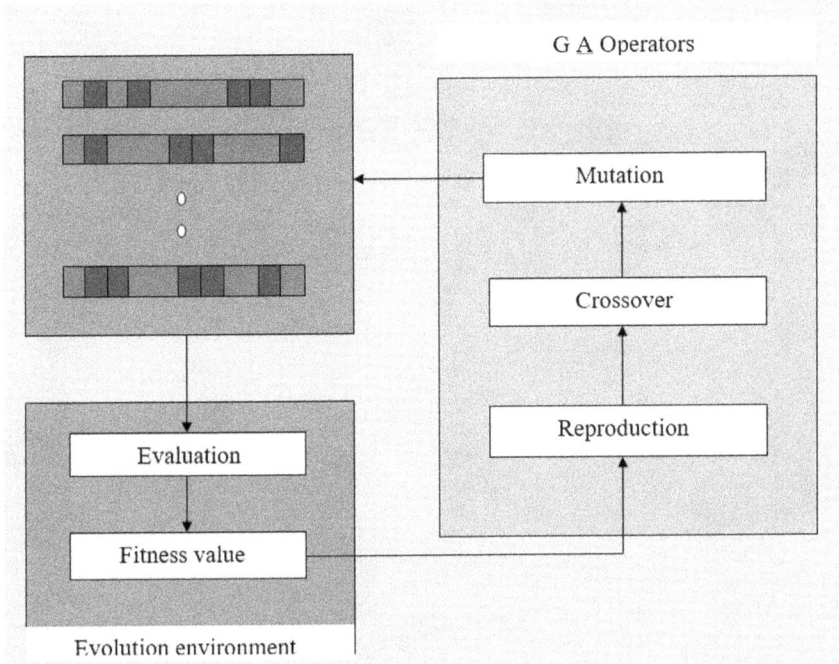

FIGURE 11.9   Flow diagram of genetic algorithm.

Step 5: mutation operation is performed on each and every chromosome.
Step 6: all the steps from Step 2 are followed until the stop criteria are fulfilled.

Some of the concepts in the above steps are described as follows:

There exists a strategy called the Russian roulette strategy as shown in Figure 11.10, which was established in order to select or elect parents for the preceding generation. For each and every chromosome the probability of getting selected to be the parent is calculated. This strategy states that the chromosome which has the higher fitness value will possess a greater possibility to be the parent because this ensures that the information will not be absent in the upcoming generation and also declares that the information of the chromosomes without better fitness value will also be passed on to the upcoming generation The next term that we came across is crossover and mutation [28–41].

First a site is chosen, then the chromosome is cut and the parts are joined to reconstruct the chromosome for the upcoming generation. The fitness value of the newly constructed chromosome is evaluated and if it is better than the parents then that is kept for the next generation otherwise the parent is chosen for the next generation [31, 33]. The most familiar way of mutation is to select a site on the chromosome and modify it using any value in a random manner. In this particular process, fitness value is not taken into account. So, here it shows that the mutation process can generate a worse chromosome but it helps in maintaining diversity [28, 33]. Figure 11.11 shows the process of crossover and mutation [44–48].

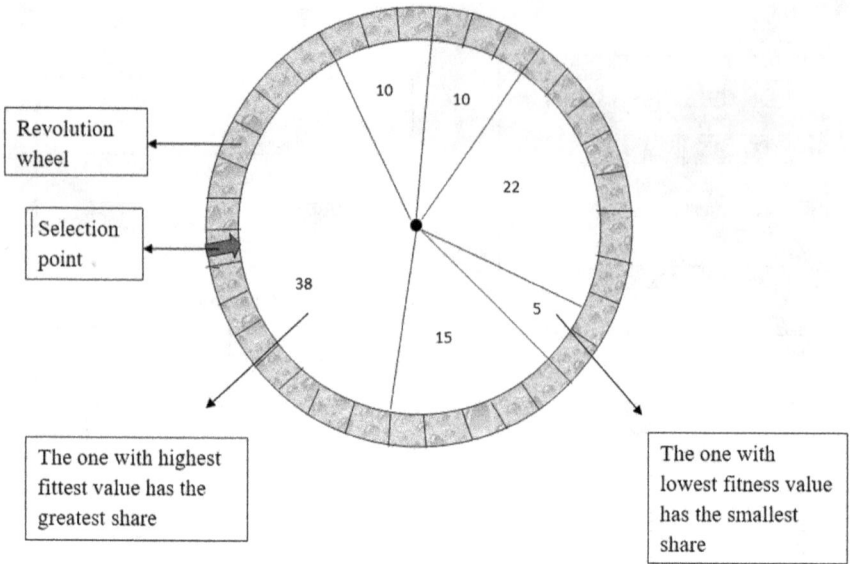

FIGURE 11.10 Russian roulette strategy.

CROSSOVER

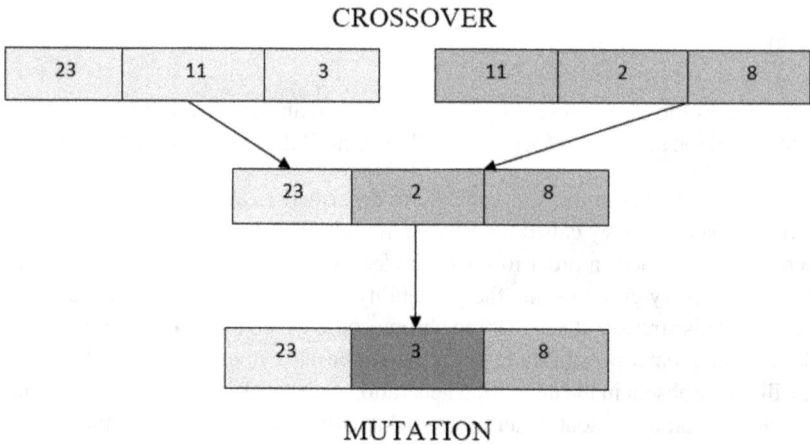

MUTATION

FIGURE 11.11 Crossover and mutation.

PSO AND NN

PSO stands for **particle swarm optimisation**. This optimisation technique was established by James Kennedy and Russell Eberhart. It repeatedly tries to enhance a candidate solution by taking into account the provided measure of quality and optimising it. Social behavior and memory are taken into account to assure that the whole population is in motion in the direction of a global optimum in a repetitious manner in the place of an evolutionary strategy. It is said to be a metaheuristic approach as it either assumes a few points or does not assume anything about being optimised and can search over a very large are of solution [18, 33].

KEY CONCEPTS OF DNN

Gradient Descent (GD)

GD is an optimisation algorithm and this represents a first order optimisation strategy. This is employed to result with local minima of a function which has can be differentiated. It is obvious that gradient represents a slope and, in such cases, the slope is estimated by taking into account the network error as well as a single weight illustrating the variation of the error with the alteration in weight. Such an optimisation strategy was first proffered through **Cauchy** in 1847. Figure 11.12 illustrates an example of GD and Figure 11.13 portrays an illustration of GD on a sequence of activation layers.

Every weight can be treated as a crucial element that will eventually lead to many transforms. The signal passes through various layers of actuation to attain the desired weight and after arriving there the relationship between the weight and error is evaluated. Chain rule is employed in this context. Chain rule associated with calculus can be depicted as follows:

$$dx/dx = (dz/dy) \cdot (dy/dx)$$

Hence, the chain rule is employed on the weight and error in following way:

- **(D Error / d Weight) = (d Error / d Activation) * (d Activation / D Weight)**
- **Error** and **weight** get attached through the **activation** layers.

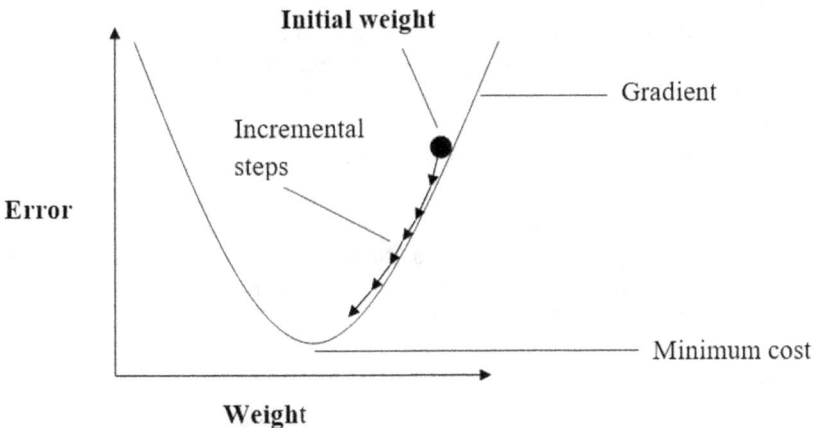

FIGURE 11.12 An example of GD.

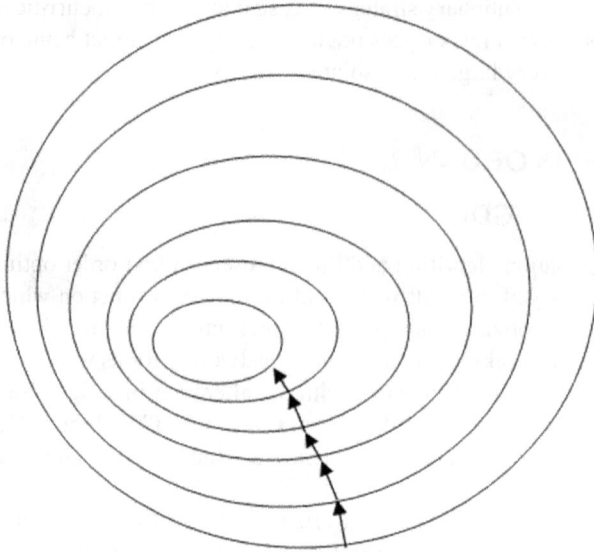

FIGURE 11.13   GD on series of activation layers input in the neurons.

- Hence, this is all about changing while balancing the weight and verifying the error generated until there is no diminishment in the concerned error.
- GD is employed to resolve and attain the linear solution as well as nonlinear equations.

OPTIMISATION ALGORITHMS (OAS)

Following are various OAs:

AdaGrad

This is a specific class of GD algorithm that adjusts and enriches the learning rate to the boundary while jointly executing small-scale alteration for the variables. For this, it handles the sparse data. This is selected to administer the sparse data. The major advantage of such a strategy is that it eliminates the necessity of changing the learning rate. AdaGrad is illustrated in Figure 11.14.

Nesterov Accelerated Gradient

This is also popularly regarded as NAG which is composed of a gradient decent step. Although it resembles a momentum approach, however, this is not fully the same. GD stage: $\theta t = yt - et \nabla f (yt)$. Figure 11.15 portrays the Nesterov accelerated gradient.

AdaDelta

The novel idea behind it is that AdaDelta resolves the drawback of AdaGrad. It rectifies the conflicting units which is there in most of the GD-based strategies. Figure 11.16 portrays a situation of AdaDelta.

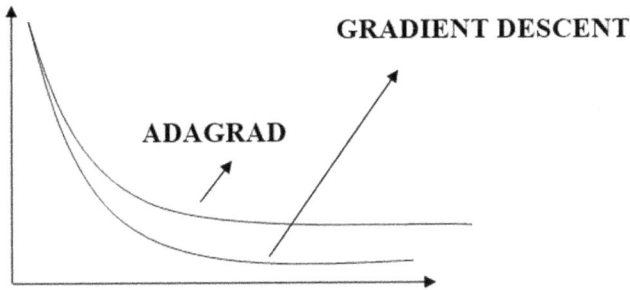

FIGURE 11.14   AdaGard and gradient descent.

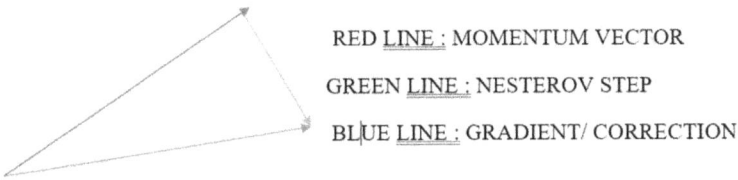

RED LINE : MOMENTUM VECTOR

GREEN LINE : NESTEROV STEP

BLUE LINE : GRADIENT/ CORRECTION

FIGURE 11.15   Nesterov accelerated gradient.

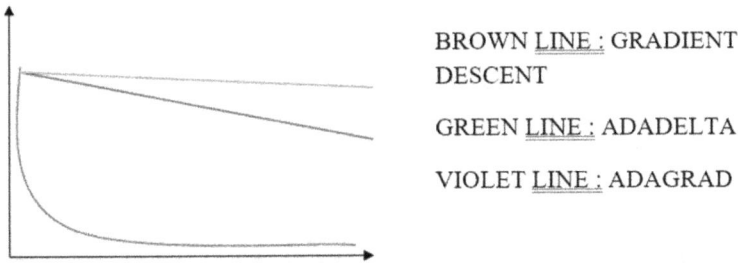

BROWN LINE : GRADIENT DESCENT

GREEN LINE : ADADELTA

VIOLET LINE : ADAGRAD

FIGURE 11.16   A scenario of AdaDelta.

Adam

To estimate the weights, in the place of a classical stochastic gradient, the Adam strategy is employed which has an advantage over the other two methods: adaptive gradient method conjointly with root mean square method. Figure 11.17 portrays the scenario of Adam optimisation in DL.

RMSProp

RMSProp is employed in training the NNs. This makes use of the average of the squared value of gradients so as to standardise it, owing to which there is fall in step size and a balance is considered, thereby restricting it to explode. Figure 11.18 portrays a scenario of big versus small learning rates. Other optimisation strategies incorporate the none approach, SGD approach, conjugate gradient strategy, Hessian free algorithm, LBFGS approach, Line GD approach.

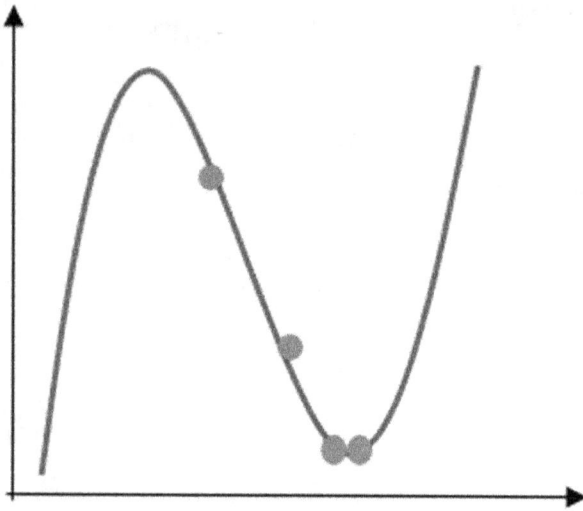

FIGURE 11.17   A scenario of Adam optimisation in DL.

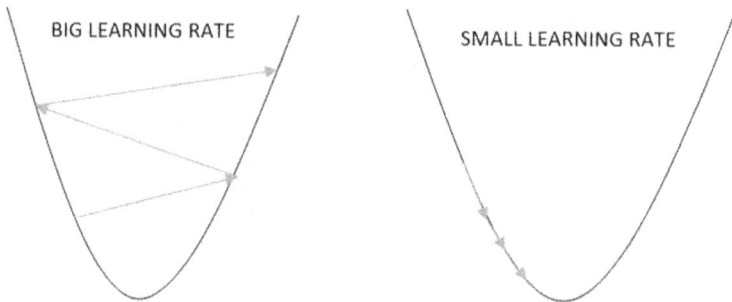

FIGURE 11.18   Big learning rate versus small learning rate.

ACTIVATION FUNCTIONS

An activation function is depicted as the equation which will assist in evaluating and regulating the output of the corresponding NN. In a NN, the data points are considered as inputs and are set in the neurons which are there in the input strata. Multiplying the number of neurons with the weight of individual layers provides the output that gets transferred to the next layer. An activation function can be treated as the pathway (as shown in Figure 11.14) between setting the input in the neurons. Figure 11.19 portrays the scenario of the activation function [5].

There are three variations of activation functions:

Binary Step Function

Such a function is designed on the basis of thresholds. Moreover, the value which is provided as the input must be lower or higher than the

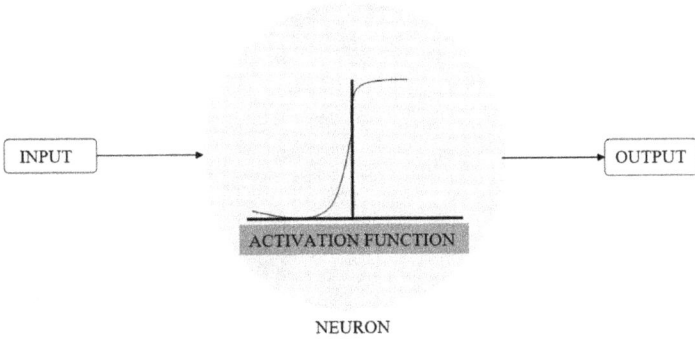

FIGURE 11.19   Representation of activation function.

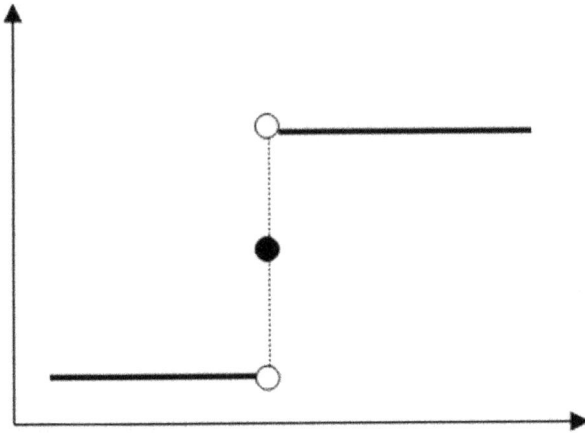

FIGURE 11.20   Binary step function.

corresponding threshold value. If this situation is satisfied then the neuron will be actuated and consequently transfer the same information or signal to the next layer. Further, this does not permit multiple outcomes. The binary step function is represented as portrayed in Figure 11.20.

Linear Activation Function

It is depicted as **A = cx** and it authorises the generation of multiple outcomes. The input value is multiplied with the neuron weight as well as the outcomes in the construction of the outcome signal. Backpropagation cannot be utilised here. An example of the linear activation function is illustrated in Figure 11.21.

Non-Linear Functions

Here, complex mapping is accomplished in between the input and output. This is heavily employed in modern NN. Any phenomenon in the NN can be depicted employing the non-linear NN. An example of a non-linear function is illustrated in Figure 11.22.

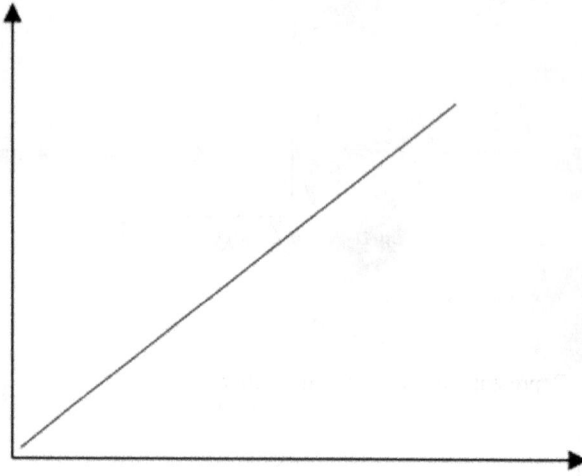

FIGURE 11.21 Linear activation function.

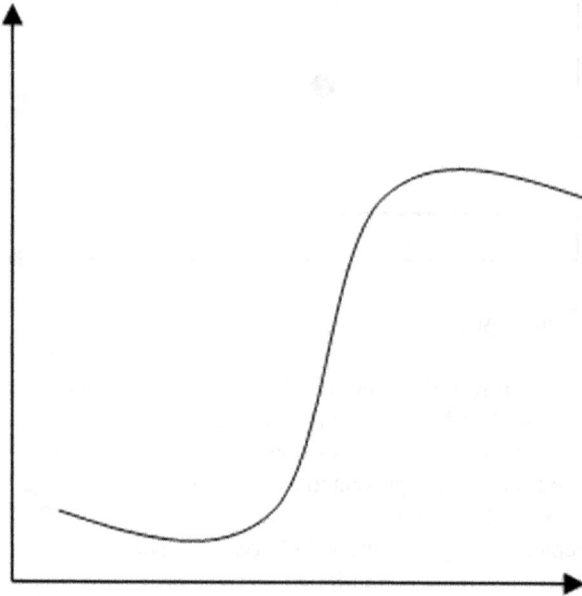

FIGURE 11.22 Non-linear function.

LOGISTIC REGRESSION (LR)

LR employs logistic functions so as to depict binary dependant variables. DNN consists of various layers that have final layers that have crucial roles. While the labelled input is attained, the output layer arranges it by employing the most applicable label.

Output nodes generate the outcomes in the form of binary values 0 or 1 [10]. This normally results in predictive evaluation. LR is employed in various domains such as medical fields, machine learning, social science, etc. Some illustrations where LR is being employed are below:

TRISS

It stands for trauma and injury severity score. It is used to estimate mortality in patients who are severely injured. Further, it can analyse and show the possibility of danger.

CONVENTIONAL OPTIMISATION APPROACHES

Optimisation approaches can be classified as follows: (A) direct search method (DSM) and (B) gradient-based method (GBM) [33].

A) **DSM**

The search is controlled through the objective function jointly with the restriction values. It is a bit time-consuming and various functions are employed to estimate for integrating and combining, and thereafter to attain the output [1, 2, 3, 46–51].

B) **GBM**

Either a first or second order derivative is employed for the 'objective function' or the restrictions so as to administer and give direction the searching phenomenon. Such an approach rapidly moves forward to generate the solution; if the concerned function and the corresponding constraints are differentiable and if they are not at all differentiable, the optimal solution cannot be attained.

The following are the drawbacks of the conventional optimisation strategy:

- The convergence is dependent on the determined first solution.
- At times, the optimal solution cannot be attained, however, we attain only the desired sub-optimal one.
- Algorithms cannot be employed on any parallel machine.
- If there is predicament concerned with the discrete search space, in such cases the algorithm cannot be employed in that situation.

CONCLUSIONS

This chapter provides a basic idea about metaheuristics, DNN, DL, applications of metaheuristics and various optimisation techniques. Our focus was to provide a clear-cut idea about DL and the related terms, and how it is implemented in real-time projects. Our main concern should be finding out the solution of various complex problems using metaheuristics and moving a step further by implementing DL in various fields. and succeeding in achieving better results.

REFERENCES

1. Gandomi, A. H. et al. (2013). *Metaheuristics applications in structures and infrastructures.* Elsevier.
2. Oussous, A., Benjelloun, F., Ait Lahcen, A. & Belfkih, S. (2018). Big data technologies: A survey. *Journal of King Saud University-Computer and Information Sciences, 30*(4), 431–448.
3. Streib, F., Zhen, Y. & Matthias, D. (2020). An introductory review of deep learning for prediction models with big data, *Frontiers in Artificial Intelligence.* https://doi.org/10.3389/frai.2020.00004
4. https://en.wikipedia.org/wiki/Activation_function
5. Zamani, M., Movahedi, M., Ebadzadeh, M. & Pedram, H. (2009). A DDoS-aware IDS model based on danger theory and mobile agents. In *Proceedings of the CiS International Conference on Computational Intelligence and Security*, pp. 516–520. Washington, DC: IEEE Publications Computer Society.
6. Nath, M. P., Priyadarshini, S. B. B., Mishra, D. & Borah, S. (2020). A Comprehensive Study of Contemporary IoT Technologies and Varied machine learning (ML) Schemes. In *Proceeding of the International Conference on Computing and Communication (IC3'20)*, pp. 623–634. Sikkim, India: Springer.
7. Nath, M. P., Priyadarshini, S. B. B. & Mishra, D. (2020, in press). A comprehensive study on security in IoT and resolving security threats using machine learning (ML). In *Proceeding of the third International Conference on Intelligent Computing and Advances in Communication (ICAC'20)*. Bhubanewar, India: Springer.
8. Nath, M. P., Priyadarshini, S. B. B. & Mishra, D. (2021, in press). Cloud computing overview of wireless sensor network (WSN). In *Proceeding of the 2nd Doctoral Symposium on Computational Intelligence (DOSCI'21)*. Lucknow, India: Springer.
9. Nath, M. P., Mohanty, S. N. & Priyadarshini, S. B. B. (2021, in press). Application of machine learning in Wireless Sensor Network. In *Proceedings of the 15th INDIACom; IEEE Conference ID 8th International Conference on "Computing for Sustainable Global Development*, p. 51348. New Delhi, India: IEEE.
10. https://machinelearningmastery.com/logistic-regression-for-machine-learning/
11. Khan, S. (2011). Rule-based network intrusion detection using genetic algorithms. *International Journal of Computers and Applications, 18*(8), 26–29.
12. Ariu, D., Tronci, R. & Giacinto, G. (2011). HMMPayl: An intrusion detection system based on hidden Markov models. *Computers and Security, 30*(4), 221–241.
13. Witten, A. H. & Frank, E. (2011). *Data mining: Practical machine learning tools and techniques* (3rd ed.). San Mateo, CA: Morgan Kaufmann Publishers.
14. Li, Y., Xia, J., Zhang, S., Yan, J., Ai, X. & Dai, K. (2012). An efficient intrusion detection system based on support vector machines and gradually feature removal method. *Expert Systems with Applications, 39*(1), 424–430.
15. Wagh, S. K., Pachghare, V. K. & Kolhe, S. R. (2013). Survey on intrusion detection system using machine learning techniques. *International Journal of Computers and Applications, 78*(16), 30–37.
16. Aghdam, M. H. & Kabiri, P. (2016). Feature selection for intrusion detection system using ant colony optimization. *International Journla of Network Security, 18*(3), 420–432.
17. Sommer, Robin & Paxson, Vern (2010). Outside the closed world: On using machine learning for network intrusion detection. In *Proceedings of the IEEE Symposium on Security and Privacy*.
18. Durakovic, B. (2017). Design of experiments application, concepts, examples: State of the art. *Periodicals of Engineering and Natural Science, 5*(3), 421–439.

19. Bengio, Y., Boulanger-Lewandowski, N. & Pascanu, R. (2013). Advances in optimizing recurrent networks. In *IEEE International Conference on Acoustics, Speech and Signal Processing*, pp. 8624–8628.
20. Begam, S., Selvachandran, G., Ngan, T. T. & Sharma, R. (2020). Similarity measure of lattice ordered multi-fuzzy soft sets based on set theoretic approach and its application in decision making. *Mathematics, 8*(8), 1255.
21. Thanh, V., Rohit, S., Raghvendra, K., Le Hoang, S., Thai, P. B., Dieu, Priyadarshini, I., Manash, S., Tuong, L. … (2020). Crime rate detection using social media of different crime locations and twitter part-of-speech tagger with brown clustering, *Journal of Intelligent and Fuzzy Systems, 38*(4) (pp. 4287–4299).
22. Nguyen, P. T., Ha, D. H., Avand, M., Jaafari, A., Nguyen, H. D., Al-Ansari, N., … Pham, B. T. (2020). Soft computing ensemble models based on logistic regression for groundwater potential mapping. *Applied Sciences, 10*(7), 24-69.
23. Jha, S., Kumar, R., Hoang Son, L., Abdel-Basset, M., Priyadarshini, I., Sharma, R. & Viet Long, H. (2019). Deep learning approach for software maintainability metrics prediction. *IEEE Access, 7*, 61840–61855.
24. Sharma, R., Kumar, R., Sharma, D. K., Priyadarshini, I., Pham, B. T., Bui, D. T. & Rai, S.. (2019). Inferring air pollution from air quality index by different geographical areas: Case study in India. *Air Quality, Atmosphere and Health, 12*(11), 1347–1357.
25. Sharma, R., Kumar, R., Singh, P. K., Raboaca, M. S. & Felseghi, R.-A. (2020). A systematic study on the analysis of the emission of CO, CO_2 and HC for four-wheelers and its impact on the sustainable ecosystem. *Sustainability, 12*(17), 6707.
26. Sharma, S., Kumar, R., Das Adhikari, J., Mohapatra, M., Sharma, R., Priyadarshini, I. & Le, D. N. (2020). Global forecasting confirmed and fatal cases of COVID-19 outbreak using autoregressive integrated moving average model. *Public Health*. doi: 10.3389/fpubh.2020.580327.
27. Malik, P. et al. (2021). *Industrial Internet of things and its applications in Industry 4.0: State-of the art, computer communication* (Vol. 166, pp. 125–139). Elsevier.
28. Sharma, R., Kumar, R., Satapathy, S. C., Al-Ansari, N., Singh, K. K., Mahapatra, R. P., … Pham, B. T. (2020). Analysis of water pollution using different physicochemical parameters: A study of Yamuna River. *Frontiers in Environmental Science, 8*. doi: 10.3389/fenvs.2020.581591, PubMed: 581591, 1–18.
29. Dansana, D., Kumar, R., Parida, A., Sharma, R., Adhikari, J. D., Van Le, H., … Pradhan, B. (2021). Using susceptible-exposed-infectious-recovered model to forecast coronavirus outbreak. *Computers, Materials and Continua, 67*(2), 1595–1612.
30. Vo, M. T., Vo, A. H., Nguyen, T., Sharma, R. & Le, T. (2021). Dealing with the class imbalance problem in the detection of fake job descriptions. *Computers, Materials and Continua, 68*(1), 521–535.
31. Sachan, S., Sharma, R. & Sehgal, A. (2021).Energy efficient scheme for better connectivity in sustainable mobile wireless sensor networks. *Sustainable Computing: Informatics and Systems, 30*. PubMed: 100504, 1–11.
32. Ghanem, S., Kanungo, P., Panda, G., Satapathy, S. C. & Sharma, R. (2021). Lane detection under artificial colored light in tunnels and on highways: An IoT-based framework for smart city infrastructure. *Complex and Intelligent Systems*. doi: 10.1007/s40747-021-00381-2.
33. Sachan, S., Sharma, R. & Sehgal, A. (2021). SINR based energy optimization schemes for 5G vehicular sensor networks. *Wireless Personal Communications*. doi: 10.1007/s11277-021-08561-6, 1–21.
34. Priyadarshini, I., Mohanty, P., Kumar, R., Sharma, R., Puri, V. & Singh, P. K. (2021). A study on the sentiments and psychology of twitter users during COVID-19 lockdown period. *Multimedia Tools and Applications*. doi: 10.1007/s11042-021-11004-w

35. Azad, C., Bhushan, B., Sharma, R., Shankar, A., Singh, K. K. & Khamparia, A. (2021). Prediction model using SMOTE, genetic algorithm and decision tree (PMSGD) for classification of diabetes mellitus. *Multimedia Systems.* doi: 10.1007/s00530-021-00817-2

36. Stolfo, S. J., Fan, W., Lee, W., Prodromidis, A. & Chan, P. K. (2000). *Cost-based modeling for fraud and intrusion detection: Results from the JAM project.* New York, NY: Department of Computer Science, Columbia University.

37. Kato, K. & Klyuev, V. (2014). An intelligent DDoS attack detection system using packet analysis and support vector machine. *International Journal of Intelligent Computitng and Cibernetics, 14*(5), 3.

38. Moustafa, J. & Slay, Nour (2015). UNSW-NB15: A comprehensive data set for network intrusion detection systems (UNSW-NB15 network data set). In *Military Communications and Information Systems Conference (MilCIS)*, Vol. 2015, pp. 1–6.

39. Moustafa, N. & Slay, J. (2015). The significant features of the UNSW-NB15 and the KDD99 data sets for network intrusion detection systems. In *4th International Workshop on Building Analysis Datasets and Gathering Experience Returns for Security (BADGERS)*, pp. 25–31.

40. Yavanoglu, O. & Aydos, M. (2017). A review on cyber security datasets for machine learning algorithms. In *IEEE International Conference on Big Data (Big Data)*, pp. 2186–2193.

41. Shiravi, I., Shiravi, H., Tavallaee, M. & Ghorbani, A. A. (2012). Toward developing a systematic approach to generate benchmark datasets for intrusion detection. *Computers and Security, 31*(3), 357–374.

42. Nweke, H. F., Teh, Y. W., Al-Garadi, M. A. & Alo, U. R. (2018). Deep learning algorithms for human activity recognition using mobile and wearable sensor networks: State of the art and research challenges. *Expert Systems with Applications.*

43. https://en.wikipedia.org/wiki/Genetic_algorithm

44. Parvat, I., Chavan, J., Kadam, S., Dev, S. & Pathak, V. (2017). A survey of deep-learning frameworks. In *International Conference on Inventive Systems and Control (ICISC)*, pp. 1–7.

45. Jia, Y. et al. (2014). Caffe: convolutional architecture for fast feature embedding. In *Proceedings of the 22nd ACM International Conference on Multimedia*, pp. 675–678.

46. Erickson, B. J., Korfiatis, P., Akkus, Z., Kline, T. & Philbrick, K. (2017). Toolkits and libraries for deep learning. *Journal of Digital Imaging, 30*(4), 400–405.

47. Anderson, J. P. (1980). *Computer security threat monitoring and surveillance* [Technical report]. Philadelphia, PA: James P. Anderson Company.

48. Heberlein, L. T., Dias, G. V., Levitt, K. N., Mukherjee, B., Wood, J. & Wolber, D. (1990). A network security monitor. In *Proceedings of the 1990 IEEE Computer Society Symposium on Research in Security and Privacy*, Oakland, CA, pp. 296–304.

49. Priyadarshini, S. B., Bagjadab, A. B., Sahu, S. & Mishra, B. K. (2018). A comprehensive review on soft computing framework. *International Journal of Advanced Mechanical Engineering, 8*, 221–228.

50. Priyadarshini, B. B. S., Bagjadab, A. B. & Mishra, B. K. (2018). The role of IoT and big data in modern technologival Arena: A comprehensive study. In *Internet of Things & Big Data Analytics for Smart Generation* (Vol. 154, pp. 13–25). Berlin, Germany: Springer.

51. Priyadarshini, B. B. (2020). A comprehensive study on architecture of neural networks and its prospects in cognitive computing. *International Journal of Synthetic Emotions, 11*, 37–55.

12 Communication Strategies for Interaction in Social Networks
A Multilingual Perspective

Bui Phu Hung and Bui Thanh Khoa

CONTENTS

INTRODUCTION

Language is defined as a means of communication, either spoken or written, in which the interlocutors engage in the encoding and decoding process. In either oral or written forms of communication, the people involved have to use sets of symbols that only those who belong to a particular language can understand. (Nunan, 2015). In other words, the message sender and recipient need to have a shared language to understand each other. One major difference between the spoken and written forms implies that oral communication usually requires the speaker and listener have a comparable knowledge and competence of the sound system in use; however, in writing, readers must use the set of written symbols encoded in the message they receive to interpret what writers mean (Nunan, 2015; Wardhaugh & Fuller, 2015).

From a sociolinguistic perspective, for effective communication, the message sender and the recipient need to have the same backgrounds. Previous views put the linguistic background of the language users as a priority for understanding. However, the contemporary literature also shows the importance of other backgrounds of the communicators, such as culture, politics, education and religion, in understanding. In a lingua franca context, no matter what level the interlocutor is at, they can use communication strategies to enhance mutual understanding (Gooskens, 2010). The recently extending scope of sociolinguistics includes one's language development due to social interaction, language variation, behaviour, multilingualism and discourse (Edwards, 2013).

Communication is crucial in the modern era; it requires a set of rules, art, manners and responses drawn through daily practical experience to communicate effectively and convincingly (Lee et al., 2020). Globalisation is no longer a new concept in the present era; however, its effects on areas of life, including communication, remains an interest in recent research (Shamne et al., 2019). Developments in technology-mediated communication (e.g., the internet, mobile phones, and social media), have provided dramatically increasing opportunities and demands for information exchange between different linguistic and cultural communities around the world. In such a context, globalisation has become an indispensable phenomenon, attracting all cultural communities worldwide. Therefore, communication between people with different linguistic and cultural backgrounds as an integral part of contemporary social life should be researched.

This chapter critically reviews aspects of communication strategies used for international communication in social networks where multilingualism and multiculturalism take place. Intelligibility takes place among people with a shared language, and their language is enhanced through social interactions. It then continues on with a discussion on the communication strategies from varied perspectives. The chapter ends with a discussion on a framework for educating communication strategies in a multilingual context.

COMMUNICATION STRATEGIES

The communication process, essentially the process of negotiating meaning, can function if, and only if, the speaker and listeners have a similar background on which their communication rests. Linguistic similarity, including grammatical structure, sound system and lexical resources, is a compulsory premise as it is the building block of meaning through which the people involved in communication understand each other via the coding and decoding process. That is, a speaker uses grammatical structure, a sound system and lexical resources of a particular language to convey his or her meaning, and only those who possess comparable knowledge and competences of that language can decode and make sense of what speakers mean. It is important to note that the role of the speaker and listener in conversations is usually interchangeable (West & Turner, 2007).

Nonetheless, effective communication requires not only linguistic competence, but also other competences: discourse competence, sociolinguistic competence and strategic competence. Communication strategies are mainly used to improve fluency and understanding in communication and to achieve communication goals.

In the context of a lingua franca, communication strategies are crucial because they can help people with different linguistic and social backgrounds understand each other. Communication strategies are what people use to convey meaning effectively in direct conversations. In a lingua franca context, interlocutors have different first and/or heritage languages. Communication strategies are usually divided according to functionality, such as asking for clarity, explaining, exemplifying and clarifying. Different competencies in the shared language show the needs for using strategies for effective communication (Wardhaugh & Fuller, 2015).

Communication strategies can also be classified into two main types: achievement and reduction. Achievement strategies enable interlocutors to take advantage of available resources to make sense of meaning because they desire to achieve their communication goals. However, the communicators may find it necessary to switch to another topic or to give up as they do not want to go on with the current meaning, so they use reduction strategies (Faerch & Kasper, 1983).

Achievement strategies can be subdivided into compensatory and retrieval. There are five main groups of compensatory strategies: code-switching strategies, interlingual transfer strategies, interlingual-based strategies, cooperative strategies and non-linguistic or non-verbal strategies. Retrieval strategies occur when the interlocutors are unable to convey a particular interlanguage item.

People communicating in a shared foreign language can switch from L2 to L1. Interactants may opt to use L1 when they have a limited linguistic resource. For instance, students with the same L1 background may use some L1 when communicating in L2. However, interlingual transfer refers to the situation when communicators use L2 with some interference of L1 both at the phonological and pragmatic level.

There are many interlingual-based strategies. The most frequently used ones are generalising, paraphrasing, restructuring, approximation, creating a new word and restructuring. L2 users may use a more general term or hypernym in place of the intended term, e.g., furniture in place of chair or electronic device in place of removable disk. Although this strategy does not demonstrate precise meaning, it improves the user's communication fluency.

Paraphrasing takes place when communicators explain their meaning with description and/or examples of the referents instead of using the relevant lexical resources. For instance, a tourist said, 'I cannot find a machine to withdraw money from my card'. The tour guide replied, 'Do you mean an ATM or a cash dispenser?' In this case, the tourist could not trace the relevant resource, so she said 'a machine to withdraw money' instead.

Restructuring refers to the situation when the speaker corrects his or her language. When delivering a presentation, a student said, 'I interviewed a couple in a park when they were sitting in a chair. Oh, sorry. I mean they were sitting in a bench'. The student first used the term 'chair' and then replaced it with the word 'bench' when she identified her word misuse.

Approximation occurs when the speaker uses an approximate synonym or another word of the same category. A speaker may use the term 'house' to mean 'flat' or 'apartment' when he or she does not have the right word in mind. Another example is the use of 'begin the engine' to mean 'start the engine'. Approximation helps the listener make approximate sense of the message.

Cooperative strategy refers to the situation when an interlocutor asks for help because he or she believes that the other interlocutor has better language competency.

For example, when the teacher is circulating the classroom during student group work, a student asks, 'Could you spell the word *psychology*?' The teacher replies, 'p-s-y-c-h and -ology meaning study'.

Speakers may also use gesture or body language to express their meaning. They may also illustrate their meaning by taking advantage of available objects or using imitate sounds. For example, a child may use sounds made by a car and gesture to mean the verb 'drive' or 'driving'.

However, speakers can simplify the language or cease the communication. Reduction strategies can be classified as formal reduction and function reduction. Formal reduction strategies occur when the speaker chooses alternatives in the lexical system to be considered more fluent and correct. Alternatively, some speakers may opt to leave the message which they want to convey or are conveying because they find it impossible to go further or explain the original message.

The communication strategies speakers develop in their first language can transfer to their second language consciously or unconsciously (Bongaerts & Poulisse, 1989). In functional linguistics, a socially oriented approach communication strategies can be classified by function, such as requesting, explaining, exemplifying, clarifying and confirming (Halliday & Matthiessen, 2014). Such communication strategies should be involved in the curriculum and are hoped to develop language learners' pragmatic competence, enhance their communication capacity and improve their confidence (Jamshidnejad, 2011).

POLITENESS IN MULTILINGUAL AND MULTICULTURAL COMMUNICATION

In the late 20th century, linguists began to be interested in communicative language, especially in communication courtesy. The term *politeness* indicates a broad category in communication, pragmatics and other cultural and sociological disciplines. There are many different definitions of politeness. Politeness is a nurturing system designed to promote successful communication by minimising the potential conflict and confrontation inherent in human communication (Lakoff, 1977). Richards and Schmidt (2013) drew on a linguistic basis to define politeness as how language reflects the social distance between communicating members and the relationships of different communicating roles. Yule (1997) argued that politeness through language is closely related to social distance or informality and formality. In short, politeness comprises communication rules in a specific culture, done through different means. All human interactions include cultural issues. If communicators comply with communication rules, mutual understanding makes communication more effective.

Brown and Levinson (1987) proposed a theory of communication politeness based on the face theory of Goffman (1981), which is considered to be the most influential in the history of studying linguistic behaviour in general, and politeness research in particular (Watts, 2003), especially in an intercultural communication environment when at least one of the two parties involved in communication uses verbal language to communicate. Leech (2016) proposed the principle of politeness;

at the same time, Kerbrat-Orecchioni's (1991) study contributed to the development of Brown and Levinson's (1987) model.

The politeness principle hypothesises that we should not take any actions or seek to mitigate the threat by utilising different linguistic means. Avoiding threatening the partner's face is a supreme principle in communication because a dialogue is a collaborative process underpinning a compromise. In a dialogue, each interlocutor seeks to contribute to the construction of negotiation or mutual understanding between parties (Kerbrat-Orecchioni, 1991). In regular communication, the parties involved are interested in protecting the partner's face to be considered polite. However, in further communications, we find that the requirement for politeness is different. If we fail to reach such a high politeness level, politeness is only considered a cliché. Communicators often pursue their goals from negotiation instead.

In human interaction, the face is the expression of one's social value, the very self-image that people need to be effectively preserved in social interaction. Brown and Levinson (1987) suggested dividing the face into two main categories, including negative face and positive face. A negative face means people do not like others to impose on their personal freedom orientation and hope not to encounter others' obstacles. Moreover, the positive face means people look forward to getting approval and love from others in the community, orienting solidarity. These two categories require the speaker to behave so that the listener feels satisfied with their principles. If the speaker can do it, they have met their face wants, and their communication behaviour is called a face-saving act (FSA). When the speaker does the opposite of the FSA, it makes the listener uncomfortable, quickly creating the possibility of communication conflict; it is a face-threatening act (FTA) (Goffman, 1981).

POLITENESS IN SOCIAL NETWORK COMMUNICATIONS

There is difference between formal and informal communication (Kouwenhoven et al., 2018). However, in linguistic research, politeness is expressed in the following three ways: (1) social distances between speakers from linguistic views, (2) influence of politeness on effective communication and (3) act of dignity (Brown, 2019).

Politeness in communication can be divided into two main dimensions: positive and negative strategies (Brown & Levinson, 1987), both of which have recently been included in communication and language research. Positive politeness may make the hearer feel pleased because their interests are satisfied; meanwhile, negative politeness strategies are oriented towards the hearer's negative face and emphasise avoidance of imposition on the hearer. Thus, the hearer makes efforts to reduce potential face threats imposed by the speaker.

Science and technology advances have provided humans with many settings for effective communication. Social networking connects people with similar interests on the internet regardless of space and time (Turban et al., 2017). Social networking has created opportunities for international connection and interlingual and intercultural communication for hundreds of millions of people worldwide to become netizens (Alshare et al., 2019). Therefore, interpersonal communication in society is not limited to direct verbal or non-verbal communication, but it is also expressed through online behaviour through social networks. Acts like hitting the reactions buttons (e.g., like, love, sad), showing emotions by using emoticons, sharing posts in various forms and commenting on posts have become common communication tools on social networks (Eginli & Tas, 2018). This communication form has created a new form of communication between online users or netizens, while it impacts previously established communication theories on direct and non-verbal communication.

THE FRAMEWORK OF POLITENESS

Humans can be defined as both social beings and conscious beings. As social beings, by all means, humans interact with other members of society, such as friends, family and other existing societal members (Dean, 2019). Besides, unlike other animals in society, humans are conscious of being; they perform rational activities or thinking when they exist (Faye, 2019).

Interaction and thinking help people realise themselves as self-concepts and are aware of other existing phenomena called other-concepts. Self-conceptualisation and other-conceptualisation processes help people find themselves and those around them as entities with shared characteristics, different individuals and social members (Wyer Jr & Srull, 2014). During the interaction process, a desire is accepted and approved as a member, cared for and shared, contributes to a positive face. They simultaneously expect that being free to do what they like, being respected, not being disturbed and not being interfered with within their private lives will create a negative face.

In the process of interacting in social networks, users have the right to communicate their views and desires to others. They use positive politeness strategies to get accepted and can be responded to by other members of the social network they are using. Social networking activities, such as comments and posts, show the users' identities. Social media users get approved by other network members on the topic of interest. However, at the same time, a social network user also wants freedom and the freedom to get what they want; therefore, they easily accept advertisements or information that they do not consider affect their lives. They can easily respond to calls from acquaintances or businesses who have purchased for sharing, commenting on posts to reduce requests for help. Furthermore, their desire to gain respect from the public may lead users to incline towards negative politeness about their posts and comments on social media. As a result, politeness, positive and negative, leads to harmony of a user's networks. From the discussion, the social courtesy model can be illustrated in Figure 12.1.

MULTILINGUALISM AND MULTICULTURALISM

Language and culture are intertwined. When a person interacts with another language, he or she also interacts with the culture embedded in that language. Similarly, he or she usually finds it hard to understand a culture absolutely if he or she does not know its language. Learning a new language not only involves the learning of the linguistic features of that language but also the learning of specific social customs, behaviours and beliefs. When learning a language, the culture to which the

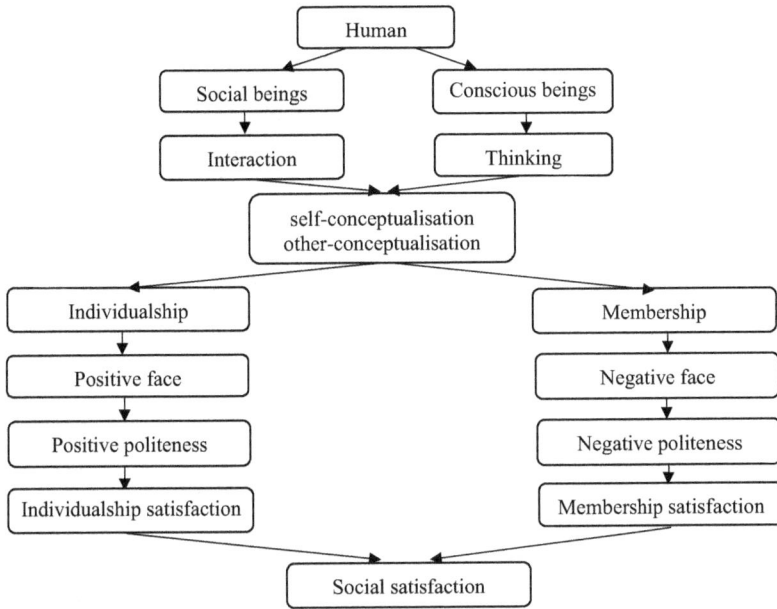

FIGURE 12.1   The politeness framework (adapted from Brown & Levin, 1978).

language belongs must be referenced. Languages and cultures develop together and influence each other as they develop. If culture is the result of human communication, communication behaviours are manifested in the culture of a particular community, which is created from interaction between the community members who use a shared language. The society into which babies are born give them their primary language and culture (Ives, 2002). As they learn, they also develop their cognitive abilities. Interactions between community members represent aspects of reality as well as connecting different contexts (Silverstein, 2004). Culture unites a community, although there is diversity in that unity. For example, the speech used by the older generation may differ from those used by young people. Furthermore, different social groups can speak the same language, but the language they use may show some subsets. The difference in the language used by an academician compared with the one used by a young clerk may be unidentical. Young netizens' posts and comments may show teen codes which their parents are not familiar with.

Globalisation and international migration also create multilingual and multicultural contexts in the world. Economic crises and armed conflicts make people move from their home country from which they have received heritage language to places where they can find employment and well-being, and then settle down in the long run. Their immediate offspring speak their heritage language, the language of the country where they are received and English. This generation is multilingual and multicultural.

Banks (2008) proposes five main concerns about multicultural education: (1) content to teach, (2) process of delivering knowledge, (3) discrimination reduction, (4) equity pedagogy and (5) school culture and social structure. Although they are considered five different areas of multicultural education, they are intersected.

Content integration refers to what the teacher uses in the classroom. For multicultural education, the materials used should have cultural diversity. The resources in language classroom can be coursebooks, visuals, teachers' linguistic features and peers' language use. The use of such resources aims to arouse learners' awareness of cultural diversity. In a multilingual and multicultural classroom, such integration may imply ethnic equality.

The knowledge construction process refers to the ways the teacher assigns tasks to students in the classroom to help them explore beliefs, cultural instances and knowledge embedded in the coursebook. The knowledge construction process does not confine itself to the classroom practice, but it includes how knowledge is introduced at the institutional level. Students' experiences through interactions on campus give them opportunities to get familiar with the paradigms of multicultural education. That is, they can conceptualise, reconceptualise and form personal knowledge and beliefs of the multicultural instances they have experienced. They can accept, reject or conditionally accept such happenstances.

Prejudice reduction helps build students' positive attitudes toward cultural diversity. Ethnocentrism is one that educators of multiculturalism are concerned with. Whether interracial contact gives students positive or negative experience partly depends on their existing attitudes. For positive experience, students should be willing to cooperate, believe in equality of humans and external interventions from teachers, parents and school authorities.

Equity pedagogy refers to the situation when the teacher's pedagogical practice aims to facilitate the learning of all ethnic groups in the classroom. The teacher should update and employ a diversity of approaches. Gay (2010, p. 29) says, 'cultural knowledge, prior experiences, frames of reference, and performance styles of ethnically diverse students to make learning encounters more relevant to and effective for them'.

Empowering school culture refers to the influence of the school culture on individuals who belong to that school. In other words, students of diverse social backgrounds can feel a sense of equality. Gaps among different groups of authorities, faculty members and students should be bridged. That a certain social group is privileged makes students aware of and feel discrimination.

MULTILINGUALISM AND LANGUAGE EDUCATION

Multilingualism demonstrates its potential to be researched. From a sociolinguistic perspective, Labov (1966) attempted to establish associations between several social issues in people's language, such as social background, race, gender and style. Other researchers (e.g., Wolfram, 1970) identified linguistic features of multilingual speakers and give implications for L2 teaching and learning in multilingual contexts.

The forms used by speakers of one social class may occur at different frequency levels from those of another social class. Their language is somewhat influenced by the community in which they have lived. This means that linguistic differences in communication between speakers of different social classes evoke linguists' research interests. As a result, speakers of minority groups may suffer disadvantages when they get involved in a larger-scale communication or communication with speakers of major groups. In some contexts, accents, vocabulary and other linguistic features frequently used by speakers of major groups are privileged, making speakers of minor groups change their own language to a certain extent if they want to meet the *language standard* determined by society. This may also occur when rural dwellers move to a metropolitan centre. However, realisations of such linguistic differences reflect the education they have received and the community in which they have grown up. In their community and education system, such linguistic features are considered *standard*. Why do they have to suffer discrimination? In an age where multilingualism is regarded as an inevitable phenomenon because of technological advances, globalisation and international communication, people need to accept language variations. The recognition of the systematicity of the language system can help testing agencies and language educators validate their measures and teaching. Language learners and test-takers, to a certain extent, show the communal values in their language. Good knowledge of such values may help test developers to truly measure what they want to measure, without any interference of other variables, and a language teacher to understand their students better.

Some experts raise the awareness that language assessment should be based on intelligibility, which refers to 'listeners' understanding, comprehensibility and accentedness'; however, this depends on the speaker's and listener's linguistic competencies and sociocultural background (Yazan, 2015). The involvement of teachers who can speak the learners' L1 may help improve the intelligibility between the teacher and learners at school. In lingua franca contexts, the teacher training curriculum should include updated pedagogy and skills to teach relevant aspects for understanding different interlocutors in communication (Jenkins, et al., 2011).

Socially oriented works with implications for language education by Halliday (1975), Leung (2005) and Vygotsky (1978) show the guidelines of competencies that the target language learners should acquire. First, learners need to have knowledge of lexical, grammatical and phonological features of the target language for effective communication. From a view of sociolinguistics, learners need to be aware of the context in which certain linguistics features should be used, including appropriateness of meaning (Nunan, 2015), language function in communication and sociocultural aspects (Street & Leung, 2010). Cohesion, coherence and unity are also other crucial aspects of a text, both spoken and written (Halliday & Matthiessen, 2014). Finally, communication strategies help interlocutors succeed in conveying persuasive messages and maintaining communication by adjusting to the context in which they are involved. This helps the speaker achieve his or her goals (Hallahan et al., 2007).

An arising concern is whether an L2 speaker should use the culture of their native language or the language in use in their international communication in which the growth of the multilingual world highlights cultural issues. We cannot force L2 speakers to apply the culture of their L2 in intercultural communication, nor do we require the native speakers to apply the culture of the L2 speakers. Educating speakers in intercultural communication may help bridge the gap between the inter-locutors (Hoa & Vien, 2019; Liddicoat & Scarino, 2013).

Language users' identity can be identified in intercultural communication. Advances in social media networks may make this type of communication easier. Discrimination, overt and covert, can restrict people to their local beliefs and values even if multilin-gual exposure is available. The language education curriculum should not only aim to develop learners' linguistic proficiency, but also educate learners' attitudes towards identity (Kobayashi, 2010; Komlosi-Ferdinand, 2020). The presence of teachers and students from other cultures in L2 education may arouse students' awareness and acceptance of multilingual matters in academic and non-academic interactions. How can interaction occur when there is discrimination between the speakers in the same context? Vygotsky (1978) introduced the sociocultural theory, which later motivates other concepts and pedagogical techniques currently applied in L2 teaching and learn-ing. The teacher may mediate students through classroom interaction (Le, 2020) by using scaffolding strategies in small groups (Van de Pol, Mercer & Volman, 2018). The classroom should be a social context in which people share knowledge via communi-cation. Proper attitudes of the interlocutors are essential for the maintenance of their communication and enhancement of the second language acquisition (SLA) process. Also motivated from socially oriented approaches, many researchers and practitioners (e.g., Han & Hiver, 2018; Nagao, 2019) suggested using genre-based L2 writing peda-gogy to develop learners' writing competence, confidence and students' awareness of text types. This pedagogy argues for classroom change into a social context in which teachers and students share knowledge and experience and give peer feedback.

CONCLUSION

Social networks make great contributions, allowing netizens to search for informa-tion and express themselves and experience life, interacting, meeting, exchanging information, connecting with the community, sharing feelings, inspiring each other searching for jobs, advertising, doing business and enjoying entertainment. The development of social networks has created a new horizon with new values, brought efficiency to the economy, and promoted education – training and other areas of social life. However, if subjective users lose direction in using social media and net-works, they may intentionally or unintentionally spread fake news and information among netizens and then even the physical world. Each user needs to have a new sense of cultural conduct to minimize negative aspects of social networks. From the practice of using social networks, a behavioural culture on the internet has been formed. Communication strategies in social networks are understood as the value system that regulates perceptions, attitudes and behaviours of individuals and com-munities in relationships with nature, society and themselves when participating. Social network behaviour reflects the development of individuals and communities. Behavioural culture in social networks includes the relationship between people and the surrounding environment, reflected in the fact that each person knows how to contribute to propagating in social networks about protecting the natural environ-ment and saving resources, living green, learning how to love, care for and protect animals. The communication strategies in the social network are also reflected in the relationship with oneself, with values such as modesty, honesty, courage, clarity, stance, the spirit of vision and learning.

The relationship between language and society has been being researched for at least a century. It is mainly interested in such topics as language variation, sensitivity, language acquisition and language transfer among different social groups through interaction. Globalisation and multilingualism have rapidly developed the world of L2 speakers.

However, in interaction, cultural differences between speakers with different social backgrounds can be identified. Language education curriculum should aim to develop learners' language competence and educate learners in cultural differences, attitudes to identity and communication strategies. Skills and self-efficacy should also be two other pivotal components of the curriculum.

REFERENCES

Alshare, K. A., Moqbel, M. & Garni, M. A. A. (2019). The impact of trust, security, and privacy on individual's use of the Internet for online shopping and social media: a multicultural study. *International Journal of Mobile Communications, 17*(5), 513–536. https://doi.org/10.1504/IJMC.2019.102082

Banks, J. A. (2008). *Introduction to multicultural education*. Boston: Pearson.

Bongaerts, T. & Poulisse, N. (1989). Communication strategies in L1 and L2: Same or different? *Applied Linguistics, 10*(3), 253–268.

Brown, P. (2019). Politeness and impoliteness. In H. Yan (Ed.), *The Oxford handbook of pragmatics* (pp. 383–399). Oxford, UK: Oxford University Press.

Brown, P. & Levinson, S. C. (1987). *Politeness: Some universals in language usage*, Vol. 4. Cambridge, UK: Cambridge University Press.

Dean, H. (2019). *Social policy*. Oxford, UK: John Wiley & Sons.

Edwards, J. (2013). *Sociolinguistics: A very short introduction*. Oxford, UK: Oxford University Press.

Eginli, A. T. & Tas, N. O. (2018). Interpersonal communication in social networking sites: An investigation in the framework of uses and gratification theory. *Online Journal of Communication and Media Technologies, 8*(2), 81–104. https://doi.org/ 10.12973/ojcmt/2355/

Faerch, F., & Kasper, G. (1983). Plans and strategies in foreign language communication. In F. Faerch & G. Kasper (Eds), *Strategies in interlanguage communication* (pp. 20–26). Harlow, UK: Longman.

Faye, J. (2019). *How matter becomes conscious*. Berlin, Germany: Springer.

Gay, G. (2010). *Culturally responsive teaching: Theory, research, and practice* (2nd ed.). New York, NY: Teachers College Press.

Goffman, E. (1981). *Forms of talk*. Philadelphia, PA: University of Pennsylvania Press.

Gooskens, C. (2010). The contribution of linguistic factors to the intelligibility of closely related languages. *Journal of Multilingual and Multicultural Development, 28*(6), 445–467. https://doi.org/10.2167/jmmd511.0

Hallahan, K., Holtzhausen, D., van Ruler, B., Verčič, D. & Sriramesh, K. (2007). Defining strategic communication. *International Journal of Strategic Communication, 1*(1), 3–35. https://doi.org/10.1080/15531180701285244

Halliday, M. A. K. (1975). *Learning how to mean: Explorations in the development of language*. London, UK: Edward Arnold.

Halliday, M. A. K. & Matthiessen, C. M. I. M. (2014). *Halliday's introduction to functional grammar*. London, UK and New York, NY: Routledge.

Han, J. & Hiver, P. (2018). Genre-based L2 writing instruction and writing-specific psychological factors: The dynamics of change. *Journal of Second Language Writing, 40,* 44–59. https://doi.org/10.1016/j.jslw.2018.03.001

Hoa, C. T. H. & Vien, T. (2019). The integration of intercultural education into teaching English: What Vietnamese teachers do and say. *International Journal of Instruction, 12*(1), 441–456. https://doi.org/10.29333/iji.2019.12129a

Ives, P. (2002). Three interventions on cultural difference, language, theory and progressive politics: A review essay. *Studies in Political Economy, 68*(1), 107–126. https://doi.org/10.1080/19187033.2002.11675193

Jamshidnejad, A. (2011). Functional approach to communication strategies: An analysis of language learners' performance in interactional discourse. *Journal of Pragmatics, 43*(15), 3757–3769. https://doi.org/10.1016/j.pragma.2011.09.017

Jenkins, J., Cogo, A. & Dewey, M. (2011). Review of developments in research into English as a lingua franca. *Language Teaching, 44*(3), 281–315. https://doi.org/10.1017/S0261444811000115

Kerbrat-Orecchioni, C. (1991). La politesse dans les interactions verbales. *Dialoganalyse III, 1,* 39–59.

Kobayashi, Y. (2010). Discriminatory attitudes toward intercultural communication in domestic and overseas contexts. *Higher Education, 59,* 323–333. https://doi.org/10.1007/s10734-009-9250-9

Komlosi-Ferdinand, F. (2020). The students, the local and the foreign: Drama of identity and language in Mongolian-English bilingual schools. *Journal of Language and Education, 6*(3), 153–166. https://doi.org/10.17323/jle.2020.10297

Kouwenhoven, H., Ernestus, M. & van Mulken, M. (2018). Communication strategy used by Spanish speakers of English in formal and informal speech. *International Journal of Bilingualism, 22*(3), 285–304. https://doi.org/10.1177/1367006916672946

Labov, W. (1966). *The social stratification of English in New York City.* Washington, DC: Center for Applied Linguistics.

Lakoff, R. (1977). What you can do with words: Politeness, pragmatics and performatives. In *Proceedings of the Texas Conference on Performatives, Presuppositions and Implicatures*, Washington, DC.

Le, P. H. H. (2020). The role of mediation in classroom interaction. In H. Lee & B. Spolsky (Eds), *Localizing global English: Asian perspectives and practices* (pp. 139–150). New York, NY: Routledge.

Lee, O. S., Lee, K. S. & Gu, H. J. (2020). Influence of perception of importance of communication, self-esteem and communication skill on patient safety attitude of nursing student. *Journal of Digital Convergence, 18*(10), 307–314. https://doi.org/10.14400/JDC.2020.18.10.307

Leech, G. N. (2016). *Principles of pragmatics.* London, UK: Routledge.

Leung, C. (2005). Convivial communication: Recontextualizing communicative competence. *International Journal of Applied Linguistics, 15*(2), 119–144. https://doi.org/10.1111/j.1473-4192.2005.00084.x

Liddicoat, A. J. & Scarino, A. (2013). *Intercultural language teaching and learning.* Oxford, UK: Wiley-Blackwell.

Nagao, A. (2019). The SFL genre-based approach to writing in EFL contexts. *Asian-Pacific Journal of Second and Foreign Language Education, 4*(6), 1–18. https://doi.org/10.118640862-019-0069-3

Nunan, D. (2015). *Teaching English to speakers of other languages.* New York, NY: Routledge.

Richards, J. C. & Schmidt, R. W. (2013). *Longman dictionary of language teaching and applied linguistics*. London, UK: Routledge.

Shamne, N. L., Milovanova, M. V. & Malushko, E. Y. (2019). Cross-cultural professional communication in the context of globalization. In *IOP Conference Series. Materials Science and Engineering* (Vol. 483, 012081. IOP Publishing. https://doi.org/10.1088/1757-899X/483/1/012081.

Silverstein, M. (2004). "Cultural" concepts and the language-culture nexus. *Current Anthropology, 45*(5), 621–652.

Street, B. & Leung, C. (2010). Sociolinguistics, language teaching, and new literacy studies. In N. H. Hornberger & S. L. McKay (Eds), *Sociolinguistics and language education: New perspectives on language and education*. Toronto, ON: Multilingual Matters.

Turban, E., Outland, J., King, D., Lee, J. K., Liang, T.-P. & Turban, D. C. (2017). *Electronic commerce: A managerial and social networks perspective*. Cham, Switzerland: Springer.

Van de Pol, J., Mercer, N. & Volman, M. (2018). Scaffolding student understanding in small-group work: Students' uptake of teacher support in subsequent small-group interaction. *Journal of Learning Sciences, 28*(2), 1–34. https://doi.org/10.1080/10508406.2018.1522258

Vygotsky, L. S. (1978). *Mind in society – The development of higher psychological processes*. London, UK: Harvard University Press.

Wardhaugh, R. & Fuller, J. M. (2015). *An introduction to sociolinguistics* (7th ed.). Oxford, UK: Wiley Blackwell.

Watts, R. J. (2003). *Politeness*. New York, NY: Cambridge University Press.

West, R. & Turner, L. H. (2007). *Introducing communication theories: Analysis and application* (3rd ed.). New York, NY: McGraw-Hill.

Wolfram, W. A. (1970). A sociolinguistic description of detroit negro speech. *Linguistic Society of America, 46*(3), 764–773. https://doi.org/10.2307/1412325.

Wyer Jr, R. S. & Srull, T. K. (2014). *Memory and cognition in its social context*. New York, NY: Psychology Press.

Yazan, B. (2015). Intelligibility. *ELT Journal, 69*(2), 202–204. https://doi.org/10.1093/elt/ccu073

Yule, G. (1997). *Pragmatics*. London, UK: Oxford University Press.

References

Index

For Product Safety Concerns and Information please contact our EU
representative GPSR@taylorandfrancis.com
Taylor & Francis Verlag GmbH, Kaufingerstraße 24, 80331 München, Germany